A PARENT'S LEADERSHIP FIELD MANUAL

Military Leadership
Principles to Help You
Lead Your Kids to a
Wonderful Life

-FAST-PACED STORIES

-VIVID PHOTOGRAPHS

MAX KLEIN

BEST WISHES TO YOU AND YOOR FAMILY! SINCERECY, [signature]

ISBN-13: Soft cover 978-0-615-34245-0
First Edition, February 2010
Printed in the United States of America
Unattributed quotations are by Max Klein
Includes bibliographical references
Copyright 2009 by Max Klein

TO CONTACT WITH QUESTIONS ABOUT ORDERS/SHIPPING, WRITE TO:

Klein's Leadership Manuals

100 Leeds Road

Newville, PA 17241

Or email: leadershipmanuals@hotmail.com

Disclaimer: This book is designed to provide information on leadership traits and their application to daily life. The purpose of this book is to educate the reader on military leadership principles and suggest how they can be applied to their own daily life. The author and the publisher and all who were involved in the production of this book shall assume no liability or responsibility for any loss or damage caused by the information and suggestions contained in this book. Furthermore, situations described in the book are based on the authors best understanding of the historical events as they happened, including the quotations listed in each chapter. If any historical inaccuracies are detected, we regret the error and we ask that you please notify the publisher (above) to request a change on any additional printings of the book.

Dedicated to:

Mom, you're still with us in spirit, and my son can feel your love through me every day. Thank you for being the most loving mother I can imagine. The memories I have of you are some of the best parts of who I am. I can never thank you enough for what you've done for me—I can only begin to show my gratitude and respect and love for you by trying to give my kids the same gifts of love, happiness, wonder, enthusiasm, adventure, and faith that you gave to me. Your goal in life was to glorify God, and there's no question you've attained that goal and are still doing so.

Dad, I've always wanted to be like you and still do. You are my main role model for being a man and a father. Thanks for being a great one.

My wife Kamm, thank you for being the wonderful person you are and such incredible mother to our son. He loves you so much and so do I. I am so happy to go through this wonderful adventure of life with you by my side.

And finally, for Cole and his little sibling on the way, I love you more than you can ever imagine. My top priority in life is to be a good dad for you. I'll do my very best! <u>I love you</u>.

Special thanks to:

My editor, Lisa Howard (www.poignantpen.com). You truly have a wonderful gift for taking the thoughts and intentions of a writer and structuring them in a way that is the most...well...poignant! You capture, distill, organize, and then amplify my thoughts and intentions in a way that I could never have done myself. Your competence and genuine caring are always evident throughout the process, from conception to completion. Thanks...it's been great working with you, and I hope you'll be my editor on the next one!

Also thanks to my little sister, Allison, for supporting me with love and professional competence in this project, and thanks to her husband, Rob Dutrey, for helping with the quotations.

Table of Contents

CHAPTER 1

"Love In Our Bloodlines"

Our Mission Defined

"When I was a young man, I wanted to change the world, so I tried to change my nation. When I found I couldn't change the nation, I began to focus on my town. I couldn't change the town; as an older man, I tried to change my family.

Now, as an old man, I realize the only thing I can change is myself, and suddenly I realized that if long ago I had changed myself, I could have made an impact on my family. My family and I could have made an impact on my town. Their impact could have changed the nation and I could have indeed changed the world."- Unknown (often attributed to a 12th century monk)

December 2nd, 1859 (Charlestown, Virginia): My great-great-great-grandpa Brown walked up the wooden steps. His defiant blue gaze fell on the sea of faces and landed on the rolling Appalachian hills surrounding him; hills whose every nook and cranny he could read as easily as a book. The breeze shifted, stirring his hair and clothes, but his body remained still. His mind was as clear and bright as the winter sun beating down on his 59-year-old skin.

Although it was a winter day, it felt like a day in early spring. It was the kind of April-like day that held the scent of melting snow and wet mud and the sound of ice dripping on a tin roof. The cool breeze brushed across Grandpa Brown's face and made his eyes water. If he'd been a sentimental man, he might have allowed a tear to fall when he'd realized that he was seeing his Appalachians for the last time, but his soul still burned with the conviction that had led him to mount the creaking wooden steps. That conviction had over-shadowed any sentimental feelings he'd had a long time ago.

Grandpa Brown paid no attention to the material being placed over his head; instead, his thoughts turned to the nearness of Death. The breeze tugged at the fabric and threatened to tear it away from his eyes. His ears heard the sheriff ask if anyone had a pin to fasten the cloth in place.

The old man didn't say a word—he simply lifted his bound arms to show the sheriff the pins his wife had tucked into the collar of his coat. Soon, the fabric was secured. He stood and waited for the enemy to act.

The sheriff asked Grandpa if he wanted to drop a handkerchief to signal when he was ready to go.

"No, I don't care," Grandpa said in a calm voice. "I don't want you to keep me waiting unnecessarily."[1] He stiffened his body in anticipation of the violent motion he knew was coming.

A few seconds later, the floor of the platform gave way with a crash. The body of my old Grandpa John Brown dropped sharply and the rope tightened, crushing out the last few seconds of his life. Only a few gasps ran through the crowd.

John Brown was considered to be both a criminal and a saint. The criminal charge brought against him was for treason against the state of Virginia for leading armed insurrections against the U.S. Government as part of his fight against slavery. This is when he and his gang of 19 men had raided a U.S. arsenal in Harper's Ferry, Virginia, and had killed seven people in the process. That had been the last straw. The government had sent in Robert E. Lee—and U.S. Marines—to capture Grandpa Brown.

Grandpa Brown had been an abolitionist who was aggressively opposed to slavery. He is one of the most notable abolitionists in history. He wasn't a violent man by nature, but his principles had led him down that road as a very last resort. He was once heard to have said, "I, John Brown, am now quite certain that the crimes of this guilty land will never be purged away but with blood." And he had the courage to *take action* to defend the principle that all men should be free. He had described the more peaceful abolitionist movement of the time when he said, "These men are all talk. What we need is action - action!" [2] His conviction to fight and die for what was right planted some of the most powerful seeds of inspiration that would sprout a few years later and begin the Civil War.

Was John Brown a military leader? Not in the traditional sense. Was he a leader who had many of the traits taught in military leadership? Yes. And he was also part of my family. That is why I have illustrated his story; a story of leadership principles in an individual having a positive effect generations into the future not only on their own family but on the world. Whether his methods were right or wrong is for you to decide, but there's no question he was a leader. The best way to illustrate the leadership traits that had enabled John Brown, and currently enables millions of others to change the world for the better, is by telling stories of life in the military where these traits are practiced daily.

July 19th, 1983 (124 years later in Northeastern Pennsylvania) I laid my head down on the prickly grass so I could breathe in the rich, sweet musty aroma of the earth. The grass tickled my nose. I rolled over again, saw a patch of yellow, and thought about how the milk from the dandelion stem was like flower's blood. There was beauty everywhere.

I was six years old and playing out in our yard on a lazy summer day. The warm-but-crisp Pocono Mountain air danced around the big Victorian style farmhouse where I lived. I grabbed one side of the thin woven green picnic blanket that we'd spread out for lunch earlier and kept rolling in it until I was engulfed in the soft cotton warmth of childhood, my mind full with innocence and freedom and wonder. Nothing else existed.

I lived in a safe country, a just country, built by the parents of the past. I was happy. Later that evening, I laid in bed and looked at the way the warm glow of the desk lamp on the nightstand shaded the corner of the bedroom orange; the tongue-and-groove satin-finished wooden wall reflected the color and deepened it into almost an amber hue. I pulled the fresh sheet around my shoulders as my mother leaned down to kiss me on my forehead and wish me good dreams. I was happy and safe with a world of adventure right outside my door and more waiting for me in my dreams. With a smile, I closed my eyes and drifted off to sleep, already dreaming of tomorrow.

> "Your descendants shall gather your fruits."
> – Virgil, (37 B.C.)

Much of my happiness was thanks to who my Grandpa Brown had chosen to be over 150 years ago and even more of it is due to the leadership of more recent relatives like my own parents, grandparents and great-grandparents. So how can we, too, know that our lives will echo love and happiness for our family into the future?

The answer is simple: by becoming *leaders of character*. Grandpa Brown was a leader who lived a strong life, a life of principle, and had the courage to fight for those principles. The choices he and many of my ancestors and current family have made in their lives have changed the future for the better, including my life now. And possibly even in a small way from reading this book, they will change yours too. You don't need to have come from a notable family or money or power. You can plant this power in your bloodlines now if it's never been there before. Your life *will* have an effect on people hundreds of years from now. The question is, what kind of effect will it have? Maybe it won't be a massive one, maybe not one that history books will be written about, but it must be a good one. The world needs to be a better place, even if just a little bit, because you *lived with character*.

So, how do we become *leaders of character* in order to have this positive effect on the future? We become leaders by understanding leadership principles and applying them to our lives. We don't have to be military trained to learn and use these concepts. Grandpa Brown wasn't in the military. Leadership principles are universal—despite the fact that they are taught most effectively in the U.S. military, anyone can learn them. If these principles are applied to parenting, they are extremely powerful and can produce incredible immediate results that will last for generations to come.

This book does not get into the nuts and bolts of specific parenting situations or teach you how to be some drill sergeant (no one wants that for their kids)—rather, this book shows you, the parent, how to **be a better leader.** Questions about specific situations will answer themselves if you become a good leader first.

In this book, you'll learn leadership principles through anecdotes, stories, and quotations from military life. All those examples illustrate how easy it can be to apply these principles to your own life.

We can only lead our children toward their potential if we are on the way to reaching ours—not in terms of wealth, status, or education, but in terms of the quality of people we are and the strength of our character. That strength of character is simply the sum of our applied leadership traits.

We probably won't need to die for our principles the way John Brown did, but we need to be willing to stand up for them no matter what comes along. Adversity, blame, lack of confidence, fear, misguided priorities, ignorance, selfishness, giving up, failure, boredom...all of these are beatable or at least more easily handled with sound leadership. The influence of your parenting does not end in the confines of your home; in some way it reaches throughout the world and far into the future. We will never be perfect parents and should never want to be, but we can always be better people and more effective leaders. We can't always change our kids and our situations, but we can change ourselves anytime. The culmination of individual parenting efforts like yours and mine determine the happiness or the misery of future humanity. It starts with us; we must change ourselves before we can change our families, and before we can change the world...and changing ourselves for the better is much easier when we truly understand *leadership*. So read on! Leadership is a great thing that *will* change your life. Think back and you'll realize that somehow, it already has.

John Brown

John Brown's Body

sung to the tune of that later inspired the writing of
The Battle Hymn of the Republic:
("Mine eyes have seen the glory...")

Old John Brown's body lies moldering in the grave,
While weep the sons of bondage whom he ventured all to save;
But tho he lost his life while struggling for the slave,
His soul is marching on.

John Brown was a hero, undaunted, true and brave,
And Kansas knows his valor when he fought her rights to save;
Now, tho the grass grows green above his grave,
His soul is marching on.

He captured Harper's Ferry, with his nineteen men so few,
And frightened "Old Virginny" till she trembled thru and thru;
They hung him for a traitor, themselves the traitor crew,
But his soul is marching on.

John Brown was John the Baptist of the Christ we are to see,
Christ who of the bondmen shall the Liberator be,
And soon thruout the Sunny South the slaves shall all be free,
For his soul is marching on.

The conflict that he heralded he looks from heaven to view,
On the army of the Union with its flag red, white and blue.
And heaven shall ring with anthems o'er the deed they mean to do,
For his soul is marching on.

Ye soldiers of Freedom, then strike, while strike ye may,
The death blow of oppression in a better time and way,
For the dawn of old John Brown has brightened into day,
And his soul is marching on.

-This version is by William Patton, 1861

CHAPTER 2

"Follow Me!"

The Fourteen Leadership Traits

First Marine Division sign posted in the sand of northern Kuwait a few days before the invasion of Iraq.

Somehow, the Humvee's padded seat, vinyl door, and steering wheel allowed just enough room to prop up a knee, extend a leg to the left of the brake, and rest a Kevlar helmet on the dusty console. It almost felt like you were lying down.

We were waiting. Waiting in northern Kuwait just south of Iraq, waiting with our vehicles staged in lines and our radios set to the same frequency and oriented north toward the unknown. We were waiting and facing unknown military resistance, unknown potential chemical weapons, unknown situations, and an unknown future for all of us—our families, ourselves, and the world.

We had been waiting for about seven hours in this line (and many more hours in other lines), but suddenly I could sense history approaching. Somehow, soon, the world was about to change. That change would be felt by those in this line of vehicles and the other lines like it that were scattered across the dark, quiet Kuwaiti desert.

I thought about the single sheet of paper someone had handed me earlier that day. It had been a message from our Commanding General, Major General James N. Mattis:

1st Marine Division (REIN) Commanding General's Message to All Hands

March 2003

For decades, Saddam Hussein has tortured, imprisoned, raped and murdered the Iraqi people; invaded neighboring countries without provocation; and threatened the world with weapons of mass destruction. The time has come to end his reign of terror. On your young shoulders rest the hopes of mankind. When I give you the word, together we will cross the Line of Departure, close with those forces that choose to fight, and destroy them. Our fight is not with the Iraqi people, nor is it with members of the Iraqi army who choose to surrender. While we will move swiftly and aggressively against those who resist, we will treat all others with decency, demonstrating chivalry and soldierly compassion for people who have endured a lifetime under Saddam's oppression.

Chemical attack, treachery, and use of the innocent as human shields can be expected, as can other unethical tactics. Take it all in stride. Be the hunter, not the hunted: never allow your unit to be caught with its guard down. Use good judgment and act in the best interests of our Nation.

You are part of the world's most feared and trusted force. Engage your brain before you engage your weapon. Share your courage with each other as we enter the uncertain terrain north of the Line of Departure. Keep faith in your comrades on your left and right and Marine Air overhead. Fight with a happy heart and strong spirit.

For the mission's sake, our country's sake, and the sake of the men who carried the Division's colors in past battles — *who fought for life and never lost their nerve* — carry out your mission and *keep your honor clean*. Demonstrate to the world there is "No Better Friend, No Worse Enemy" than a U.S. Marine.

J.N. Mattis, Major General, U.S. Marines, Commanding

Although this message came directly from our commanding General—our leader in Operation Iraqi Freedom—he was just one of many leaders who made our mission successful. In a few words, he painted a clear picture of our mission, our responsibilities, and his confidence in our abilities...and added an extra spark of motivation that our General was right there with us and believed in us. That's what a leader does.

But leadership is not only for Generals, nor does it reside only in words. Leadership is to be practiced by everyone and in every medium, including parenting. Leadership is the most valued and revered and practiced quality in the entire military. Without leadership, the military would not function, just as a family without leadership becomes dysfunctional.

Leadership starts in Boot Camp or Officer Candidates School with the basics: first, we must learn to follow, because good followers make good leaders. In Boot Camp, one of the first things we memorized was the list of the 14 leadership traits. Their acronym is JJ DID TIE BUCKLE.

Though no one ever fully masters *all* of the traits of leadership, the principles can be quickly learned and sometimes easily implemented, but also require continuous development through a lifetime. This following list is as meaningful and useful to a child as it is to a General.

So here they are—these traits are the backbone, the foundation, the basis of all leadership development to come:

JUSTICE is defined as the practice of being fair and consistent. A just person considers each aspect of a situation and then rewards or punishes accordingly.

JUDGMENT is the ability to think about things clearly, calmly, and in an orderly fashion so that you can make good decisions.

DEPENDABILITY means that you can be relied upon to perform your duties properly. It means that you can be trusted to complete a job. It is the willing and voluntary support of the policies and orders of the chain of command. Dependability also means consistently putting forth your best efforts in an attempt to achieve the highest standards of performance.

INITIATIVE is taking action even though you haven't been given orders. It means meeting new and unexpected situations with prompt action. It includes being resourceful when something needs to be done and the normal material or methods are not available to you.

DECISIVENESS means that you are able to make good decisions without delay. Get all the facts and weigh them against each other. By acting calmly and quickly, you should arrive at a sound decision. You announce your decisions in a clear, firm, professional manner.

TACT means that you can deal with people in a manner that will maintain good relations and avoid problems. It means that you are polite, calm, and firm.

INTEGRITY means that you are honest and truthful in what you say or do. You put honesty, sense of duty, and sound moral principles above all else.

ENDURANCE is mental and physical stamina. It is measured by your ability to withstand pain, fatigue, stress, and hardship. Example: enduring pain during a conditioning march in order to improve stamina is crucial in the development of leadership.

BEARING is the way you conduct and carry yourself. Your manner should reflect alertness, competence, confidence, and control.

UNSELFISHNESS means that you avoid making yourself comfortable at the expense of others. Be considerate of others. Give credit to those who deserve it.

COURAGE is what allows you to remain calm in the face of fear. Moral courage means having the inner strength to stand up for what is right and to accept blame when something is your fault. Physical courage means that you can continue to function effectively when physical danger is present.

KNOWLEDGE is the understanding of a science or art. Knowledge means that you have acquired information and that you understand people. Your knowledge should be broad—in addition to knowing your job, you should know your unit's policies and keep up with current events.

LOYALTY means that you are devoted to your country, the Corps, and to your seniors, peers, and subordinates. The motto of our Corps is *Semper Fidelis!* ("Always faithful.") You owe unwavering loyalty up and down the chain of command, to seniors, subordinates, and peers.

ENTHUSIASM is defined as a sincere interest and exuberance in the performance of your duties. If you are enthusiastic, you are optimistic, cheerful, and willing to accept the challenges. (USMC.MIL, 2003).

Most of us are naturally strong in a few (or maybe a majority) of these traits, but there is ALWAYS room to develop them and we must develop all of them. We need to have all of these traits, because even if we're lacking just a few, all the others can become meaningless, or much worse, dangerous. For example, if you don't have the trait of courage, all of the other traits become impotent and they cannot be put into practice. If someone has knowledge without courage, what they know is irrelevant if it's never used to help anyone in a time of crisis. A knowledgeable coward is not leader. If they had courage without integrity, a life of crime would be an attractive pursuit. Another example is a far worse scenario; using the traits for the advancement of evil. Adolf Hitler was incredibly strong in all of these traits except for two of them and these two made all the difference. He completely lacked justice. Nothing about his morality and actions towards people that were physically different was just. He also lacked the integrity to the moral truth that ethnic cleansing is wrong. We could come up with scenarios like this for all of these traits. The point is that we will be strong in some traits and weak in others, but in order to become a well rounded leader, we need to at least have all of these traits and work to improve them continuously throughout our lives, especially the weakest ones.

These traits are not listed in order of importance, but in my opinion integrity and courage are the most critical ones to develop. As you work on them, you decide for yourself. These fourteen traits are the backbone of leadership, they are the fundamentals.

We are required to understand them, then put our own twist on them because no two good leaders have the same style, but all good leaders share similar fundamentals.

As you develop these traits yourself, you will be able to cultivate them in your kids. Remember that everyone is at a different level of leadership competence and everyone has different levels of potential. We can't get into the trap of comparing ourselves to others when progressing with these traits, but rather compare yourself to your potential. In other words, being as good as you can be is what counts, not how you stack up to other parents. When we take these traits to heart and begin to incorporate them into our actions, we become a good example to our kids. They begin to understand the fundamentals of leadership themselves just by observing us.

A universal truth in parenting is that your kids want to be like you (even if they don't show it when they're teenagers). Becoming a leader of character ensures that if your kids reach their goal of being like you, that'll be a good thing! Evaluate yourself honestly and often and try to work on the traits you know you are lacking. You're already doing a wonderful job because you love your kids so the further development of these traits will just be the icing that makes your life sweeter. A little development here and there is all it takes, so post the list of traits on your wall or put a list in your wallet (see cutouts in back of book). By adopting the traits for ourselves and putting them into practice, we can strengthen our leadership skills and strengthen the example we set. This is one of the greatest gifts we can give. And remember, you can become a leader of character at any time in your life. It doesn't matter what you've done in the past, how old you are, or how weak or strong you are on certain traits. Anyone can improve at any time, no matter what.

Start putting these traits into practice today. You'll be amazed at the results!

> "I hope I shall always possess firmness and virtue
> enough to maintain what I consider the most enviable
> of all titles, the character of an honest man."
> – George Washington

I laid back in my seat and listened to the Kuwaiti wind pelt sand into the canvas door and make its way across the twilight desert. Most people across the globe must have had their minds on that particular spot in the world at that particular time, I thought. The stirring sense of being on the cusp of history (and on the right side of it) heightened my senses and made me realize how glad I was to be there.

We all sensed that the waiting we'd been doing for the past few months—ever since we'd come to Kuwait—was coming to an end.

"We're bombing Baghdad!" someone yelled.

And soon after, we were heading off into the dark night towards the border. Leadership ability saturated the ranks from Private to General. Every person had learned what leadership was and had been developing those traits (that you now know) for years, so we knew we could handle what was to come. We were mentally and physically ready and we trusted those around us because we knew that they were ready, too. Strong leadership ensures that kind of confidence and optimism within the ranks in the military. Likewise, your strong leadership will inspire that confidence and optimism in yourself and in your children as you all face the unknown future together.

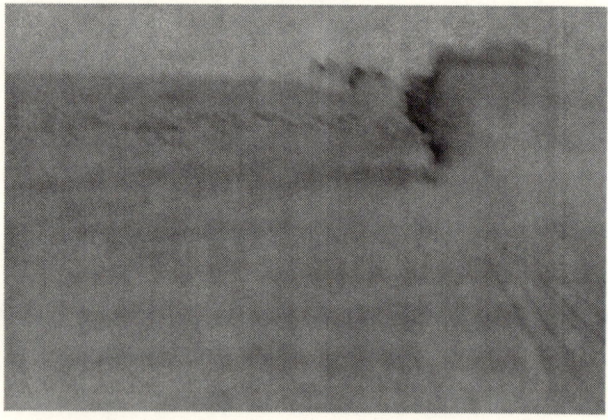

Our mission starts (1st Marine Regiment elements approaching the Kuwait/Iraq border).

"By profession I am a soldier, and I take pride in that fact. But I am prouder – infinitely prouder – to be a father. A soldier destroys in order to build; the father only builds, never destroys. The one has the potentiality of death; the other embodies creation and life. And while the hordes of death are mighty, the battalions of life are mightier still. It is my hope that my son, when I am gone, will remember me not from the battlefield but in the home repeating with him our simple daily prayer, 'Our Father who art in Heaven.'" – General Douglas MacArthur

Crossing the sand pit and berm separating Kuwait (on left), the "known," from Iraq (on right), the "unknown." We crossed a bridge built by our combat engineers.

You will help engineer with *your* leadership similar bridges of confidence and optimism with your children from the known present to the unknown future and will avoid that pit of fear that might stop others.

CHAPTER 3

"Mission Accepted"

Taking Responsibility

"Man must cease attributing his problems to his environment,
and learn again to exercise his will - his personal responsibility."
– Albert Schweitzer

There it was, hanging all alone on the end of the bunk. My breathing was heavy and I kept hearing the same words repeating in my ear.

"Don't be the one, recruit! Don't be the one who shuts down the whole base because you lost a weapon!" the drill instructor had yelled as the veins practically popped out of his neck. "Find it!" had been the final command I heard as I sprinted across the sandy, flat field toward the barracks. When a weapon is lost on a military base the roads into and out of the base are closed, all people (families on base, civilians, military personnel) are not allowed to go off base and no one is allowed on until the weapon is found. It's a big deal to say the least and you do NOT want to be the one that loses a weapon.

I had been named the Platoon Guide in the early weeks of Boot Camp and had managed to hang onto that role through rifle range week...which was midway through the duration of the entire Boot Camp. As

I ran, my scattered thoughts kept touching on the fact that somehow I was still Platoon Guide…at least, until now. I ran into the squad bay and saw the weapon hanging on the bunk, unlocked.

One of the other recruits had gone to sick call that morning and had therefore been required to leave his M16 A2 service rifle in our—his—platoon's custody. He had apparently told his bunkmate to watch over it, but in the hellish chaos of a usual Boot Camp morning, the bunkmate had forgotten about it, and we hadn't taken the weapon with us to the range. In the end, as the Guide, I was the one responsible for it.

Minutes before—during the rifle count at the range—we'd realized it was missing. We counted again and again. Sixty-five, 66…where's 67??. A cold sweat had begun to trickle down my cheek.

A few minutes later, I was standing and looking at the weapon. I grabbed it, thankful it was still there so at least the entire base wouldn't have to be locked down, and ran back across the field to the drill instructor. I knew that I'd be "fired" as Platoon Guide.

"Sir, this recruit [you talk in the third person in Boot Camp] found the weapon and confirmed that it was the missing weapon. This recruit should have ensured weapon count prior to leaving the squad bay." I could have blamed the recruit who had forgotten to tell his rack mate that he was going to sick call, but I didn't—that was irrelevant. I had been assigned the task of ensuring that all weapons were accounted for. I had failed.

The drill instructor grabbed the weapon out of my hand the way an angry bear would maul a sapling. "You will not make that mistake again! You understand, Recruit Klein?"

"Sir, yes, sir!" I shouted. I knew I'd just been "fired."

> "The buck stops here."
> – A sign on the oval office desk of
> President Harry S. Truman
> declaring his personal responsibility.

Sometimes when times get tough and we make mistakes or when things just go wrong in life, it's much easier to blame someone or something—doing so temporarily (and sometimes permanently) takes the heat off of us so that we don't have to suffer the consequences of our mistakes. But this is how an awful disease starts inside of us *and* how an awful example is set for children.

The disease is called "victimization." Side effects include settling for failure since, "hey, it's not my fault," having a low sense of self-worth, and experiencing deteriorating morality, sadness, and feeling that life cannot improve on a financial or personal level. Victimization also spawns lies, because lies are how the victim justifies his or her situation. And implying that we're a victim of our own natures by saying "it's just who I am" is not a viable excuse, either, seeing as we're given the free will to improve who we are.

This victimization disease eventually kills the soul by making a person DISHONORABLE. Self-made victims blame schools or teachers for their children's bad behavior or bad grades. They blame the wealthy for keeping them poor, blame the boss for firing them, and blame society—or their childhood, or their history, or *other people*—for making them unhappy. Now, there are cases of real victimization, and there are instances wherein people need real help. These cases need to be tended to with extreme care and support, and if it's you, who needs help, don't be ashamed to ask for it. This chapter is not meant to diminish those who are victims of crime, rape, serious oppression, etc. What this chapter is addressing is the pseudo-victim mentality, that culture of excuses that many people adopt when they don't have the guts to take responsibility for improving their own lives.

This pseudo-victimization is dishonorable and is the antithesis of taking responsibility. We must start the process of taking responsibility by forgiving anyone or anything that has done us wrong, thus erasing them as a cause for our problems. Blame or non-forgiveness is a quick and easy way out, but in the long run, blaming outside influences leads to sadness, despair, and dishonor. That kind of mindset needs to be nipped in the bud in childhood before it can grow into an adulthood monster and remain part of a family for generations to come. We must never blame others in front of our kids in ANY situation. The bitter, unhappy, discouraged mentality of a *victim* is like a parasite that crawls into the mind and feeds off a bitter, vengeful bloodline for generation after generation.

The opposite of this hellish blaming-victimization scenario is *taking responsibility*. This is always hard to do at first and is often a shock to our system. Taking responsibility is much harder than blaming someone else, because taking responsibility means taking all the heat and fixing the problem. It means resigning yourself to working hard to fix it, whether that means getting a better job, being a better person, stopping doing drugs, or anything that is prohibiting you from approaching your potential in life. When that shock and heat wave of taking the bull by the horns dissipates, though, it leaves in its wake a little seed—a seed of honor. You may be embarrassed and in trouble for taking the heat for your mistakes, failures, and shortcomings, but you now have honor growing inside of you, and that honor doesn't leave room for the victimization parasites to thrive.

This seed of honor is watered every time you *take responsibility*. As you and this seed mature, the seed grows into a strong and powerful tree of honor that bears all of the other fruits of the good life: dignity, self-esteem, morality, truth, maturity, personal and professional advancement, and character. Most of all, this tree bears happiness. This strong oak of honor that you have chosen to cultivate over the parasite of victimization will provide comfort and shelter for you and your family amidst the cold rains of temptation and adversity. If you choose to remain a victim in any way at all, however, and you do not take responsibility for your own life and for building your honor, the good fruits cannot give sustenance to you or your family.

The first thing we parents must know is that the buck stops with us. It's no one else's fault for how our family is. *Nothing* is anyone else's fault. If you want a different situation, set about changing it, or learn to gracefully adapt to it, and don't let excuses thwart your mission. Excuses are the sound that the parasite makes when it knocks at your door and wants to slither into your soul. It's our life and our family's lives—we must take full responsibility for them. Though we can't control everything that happens to us and our family, we *can* control how we react to it...there's always an honorable way. Taking responsibility is our duty.

> "Ninety-nine percent of all failures come from people who have a habit of making excuses."
> – George Washington Carver

Later that afternoon, one of the recruits had gotten a care package of Cheese Whiz and crackers in the mail from his mother. It was promptly confiscated since only good old Marine Corps chow was approved for consumption in Boot Camp.

For the next two days, I had to march behind the platoon instead of leading it, with the colors (flag or guidon) slung over my shoulder in shame, rolled up with rubber bands; the shiny point that usually glistened in the sun was caked with Cheese Whiz.

Truth be told, I'd had a hard time not laughing at the verbal harassment I received from the drill instructors of the other platoons passing by. They'd reminded me and the rest of us that we must take responsibility for ourselves and each other, of course...but they'd said so in more colorful terms. Two days later, the Drill Instructor unfurled our flag and authorized me to remove the Cheese Whiz from the crest of it.

He pulled me aside that evening. "Recruit Klein, you took responsibility and lived up to the consequences of your error. You're still the Guide." He paused. "You understand?"

A wave of relief swept over me. "Sir, yes, sir!" I said. An unauthorized smile cracked my lips and a re-affirmation of a lesson sank in...*never blame others and never make excuses for your mistakes; take responsibility.*

I was permitted to hold the position of Platoon Guide all the way through the rest of Boot Camp. At graduation, I carried the unfurled and shiny-tipped flag in front of the platoon and all of our families and honored guests.

It was an honor I'll never forget.

"Most people do not really want freedom, because freedom involves responsibility, and most people are frightened of responsibility."
– Sigmund Freud

CHAPTER 4

"In the Arena"

Finding Out
What You're Made Of

The southern sands of Iraq. The invasion rolls north.

"A man does what he must in spite of personal consequences, in spite of obstacles and dangers, and pressures, and that is the basis of all human morality."
– John F. Kennedy

The desert was dark and quiet except for the flashes of artillery rounds exploding in the city ahead of us. The flashes looked like big green fireflies through our night vision goggles. A few seconds after we saw the flash, a "boom" would ripple by and dissipate into the empty desert behind us. An occasional hint of diesel fumes from the generators that hummed in the back-ground swirled around our parked vehicles.

It was about 5 AM when we crossed the bridge into the outskirts of An Nasiriyah in Central Iraq. The invasion had started a few days before; now, we were about half-way to Baghdad. We had driven north for about 36 hours through southern Iraq with little resistance, but it seemed as though we were finally nearing some cities where the enemy was waiting. As I drove over the bridge and looked down the other side, I saw overturned and burned-out tan Army military vehicles. U.S. vehicles! The convoy stopped at the intersection. We fanned out to deter any insurgent vehicles from entering our convoy.

"Stop that car!" someone shouted. We raised our M-16s and focused our sights on the car bomber (or oblivious civilian) ahead of us. Either he was intent on running into our convoy, or he was almost unaware we were even there until suddenly—*screeeeeeeccchhh!!*—he saw us. He must have been a civilian, because he pulled a fast U-turn and headed back to town...possibly to change his pants?

I looked at the destroyed vehicles and wondered how U.S. Army units had gotten into our area of responsibility. (A day later, I found out that a convoy had gotten lost and been ambushed. U.S. prisoners had been taken. This was the convoy, in fact, that Private First Class Jessica Lynch had been part of.)

We waited a few hours for the Infantry units to finish securing the last bridge across town so that the rest of the regiment could pass through.

With the bridges secured, we hopped back into our vehicles and drove quickly through the city and into another day. We only got potshots that day, nothing too serious. The vehicle in front of us had his tire shot out, too, but that was about it for the action in An Nasiriyah—soon, we were through the city.

It was March 24th, 2003; somewhere, a sandstorm was brewing out in the desert....

Waiting just south of An Nasiriyah, artillery explosions can be seen in the background. (I took this photo by holding my camera up to my night vision goggles.)

Many people are haunted by the fact that they've never been "tested." Many, they think, haven't had the opportunity to see what they're made of. Not in a war, per say, but under the fire of adversity or a difficult challenge or an emergency. What would I do if "X" happened? they wonder. X can be death, a car accident, home invasion, mugging, a lawsuit, jail time, a fight, witnessing a fight, or having to defend someone. This question hangs like an invisible yoke on the necks of many: can I defend myself? Can I defend my family? How would I act in a tense or emergency situation? Would I fold up, run away, or meet it with competence despite my fear?

This emotional hurdle of answering this nagging question can be overcome in many ways. The best way for children to "see what they're made of" physically is for them to participate in sports or activities that require physical courage to meet a challenge (which most sports do.) These little steps of courage result in burgeoning self esteem and confidence that they can further develop, and will have when required for real adversity. Any task—no matter how small or slight— that takes moral or physical courage to overcome is a way to "test" oneself. These tests are very, very important for children to face and overcome in order for them to develop into confident adults.

Other options for learning how to deal with stressful situations are to take classes in subjects like CPR and first aid that prepare you for emergency events. Martial arts training or weapons training might give you a reasonable shot at defending your own family should the need arise. That's a responsibility that shouldn't solely lie with law enforcement—not being able to defend our families (at least, to an extent) is shirking our duties.

It is so much more rewarding to be the person who *acts* during an emergency rather than the one who stands by and watches the situation unfold. Most of us have done both and therefore know the feelings associated with both—one is meaningful and powerful while the other is embarrassing and weak.

Marines are trained in First Aid, CPR, crisis management, and staying calm under pressure. All of this training helps to erase that question of *how would I act* and leave a sense of confidence in its wake. It is such a wonderful relief to prove to *yourself* how you'd act and thereby no longer feel the need to prove it to others. Even if you screw it up during the actual emergency, at least you did *something*.

Marines are also regularly thrust into interpersonal violence training such as pugil stick bouts and practicing martial arts. This training, too, can alleviate the "need" to prove yourself every time someone wrongs you; in other words, you're able to walk away from unnecessary confrontations. You no longer have to let others drag you down to their level and sully your character by doing so. The louder the mouth, I've noticed, the lower the confidence. The toughest people I have ever met were calm and soft-spoken at all times unless they were forced to be something else to protect someone. They are calm but capable, powerful but silent, passionate but tempered, and when the situation escalates with people like this, they become intense but controlled, and channel aggression or anger into effective action. They control their emotions and use them as a tool to solve problems, rather than letting their emotions escalate the problem. This is our goal. Are you a confident and competent presence, a tension escalator, or just an observer? People should always feel safer and more comfortable when you're around them, not just in tense situation, but in all situations. No matter how you have acted in the past, you can always improve with a little training.

If possible, train in martial arts, competitive sports, or some other challenging physical activity. Or challenge yourself in anything that requires physical courage. It's uncomfortable, but we always have to push ourselves out of our comfort zone in order to grow. Always do what you fear to do bit by bit. Nurture your confidence, grow into it. Though not the only choice for activities, martial arts or self-defense classes are invaluable. Even if you get your butt kicked, your confidence skyrockets because you're no longer weighed down by the "what if" yoke. Physical activities build confidence…which builds courage…which we need to be well-rounded leaders and parents. Setting the example of *taking action* in a critical situation is one of the most powerful examples you can set for your kids.

In a real situation, when *taking responsibility demands* that you *take action,* you will already have *tested yourself* and you will be able attack the problem with that controlled intensity that makes someone a true leader. Gaining this capability is our duty, and the feeling of true self-confidence you'll cultivate will always reside within you. You may already be confident in yourself, but if you do need more self-confidence to ensure that you *will* take action when it's needed and set that example for your kids, start training!

"Hard training, easy combat; easy training, hard combat"
– Alexander Suvorov (18th Century Russian General)

The southern outskirts of An Nasiriyah, looking west. This is where Jessica Lynch's convoy had been attacked the day before.

The next morning, the 25th of March, a sandstorm had arrived. A dim, orange glow engulfed us and sand was everywhere. On the far side of the city, bodies were strewn everywhere...and parts of them had been flung so broadly across the road that you couldn't really avoid them with your tires. The smell of death and burning clothing swirled through the orange air. It was a horrible and eerie scene, but at the same time, the smell and the view of death heightened the rich sweetness of being alive yourself and that unexpected gift will always be with me.

We pulled over as the Medevac helicopter landed.

"We need hands now!" yelled the Corpsman. It was obvious that he had an ambulance full of injured civilians.

We ran through the mud on the Euphrates River flood plain to the back of the Medical Hummer and looked inside to see blood, bandages, and IVs covering all of the bodies.

"One at a time!" the Corpsman yelled.

In the back of the Humvee, I saw a bloody Iraqi about my age who looked very much like an old high-school buddy of mine. He gave me a scared look; I tilted my head in a "you'll be all right" nod.

A litter handle was passed to me. On the litter was a small body covered in a green blanket and draped with an IV bag. We ran as quickly as possible across the mud plain towards the thumping, roaring Medevac helicopters that were waiting for us in the field.

Wind from the rotors whipped across the mud as we reached the back of the helicopter. Two Corpsman grabbed the litter and we ran out to make room for the next injured civilian. Earlier that day, the Iraqi Republican Guard soldiers had been using these innocent people (women and children) as shields by shooting at our Marines from a bus full of them. Our Marines hadn't had any choice in terms of how to react to the attack ("No Worse Enemy"), but the Iraqi soldiers had certainly had a choice to put civilians there in the first place. Regardless of the enemy's dishonorable tactics, we were making every effort to treat the innocent with every means possible; it was part of our mission. "No Better Friend."

If you have the courage to push yourself in any area of life you know you need to improve and test yourself often, you will have that confidence needed to *take action* when and where it's needed! Test yourself and train yourself and you will become strong. Soon *you* will be "no better friend, no worse enemy" like a U.S. Marine. It's a motto we can all live by as parents and a powerful leadership example for our children of how to live honorably.

> "Courage is not simply one of the virtues,
> but the form of every virtue at the testing point."
> – C. S. Lewis

CHAPTER 5

"The Color of Character"

Putting Substance Over Image

> "The qualities of a great man are vision,
> integrity, courage, understanding, the power
> of articulation, and profundity of character."
> – Dwight Eisenhower

Drops of sweat rolled down the side of my face and clung to my nose for a moment before they let go and splattered into the sandy muck.

"Straighten those backs, recruits!" the drill instructor barked. "I don't care how tired you are—I'll keep you here until *I'm* tired!"

Just focus on something else, I thought. *The more you think about how much your muscles are burning, the more they're gonna burn.* I glanced over to the recruit next to me and saw how straight his back was...and how much sweat was dripping off of his face. A little arrow of strength shot through my veins. Somehow, pain isn't as bad when it's shared.

Sixty-five of us were in a sand pit at the hot, swampy, southeast tip of South Carolina. An impassable marsh surrounded the outcropping of land. We were Platoon 3003, Kilo Company, 3rd Recruit Training

Battalion, and we were only in the third week out of *thirteen* weeks of Boot Camp. Parris Island is the training grounds of the Marine Corps Recruit Depot—grounds that are both dreaded and revered. (Dreaded before you arrive there as a recruit, but revered after you have earned the title of United States Marine).

Seems like we've been here all day, I thought, frustrated, and felt the sand fleas crawl over the back of my neck and into my ear. Bite...crawl... bite...crawl...they seemed to exist only to make things more difficult for us recruits.

"You *don't move* at the position of attention, re-cruits—you understand that? —discipline!" the drill instructor shouted as we burned and itched and soaked the sandy soil with our sweat.

"SIR-YES-SIR!" we shouted back in unison. We knew what he meant—there's never an excuse to disobey a direct order to stand at attention. Never. We were training that afternoon in close-order-drill, or marching. Sometimes we'd stand at attention for fifteen or twenty minutes, not moving a muscle or twitching an eye, just to work on self-discipline. The sand fleas chewing on one of the recruit's eardrums had been too much for him and he swatted at it, but he wasn't the only one. The self-discipline we had lacked—a shortcoming that had led us to move when we never should have—is what had put us in the sand pit. Sand fleas are no excuse to disobey an order, and we were paying the price.

"Are you ready to march back to the parade deck and show some self-discipline?" The drill instructor's voice was whip-sharp; we sensed we were about to make good on what we owed.

"SIR-YES-SIR!" Hope made our voices even louder.

"Well, I don't think you are—this recruit over here thinks I can't see him resting on his knees!"

A small, almost-inaudible groan rumbled through our ranks...or maybe the sound was just in my head, seeing as being dumb enough to voice our displeasure would lead to more of it.

The levels of building frustration and stress were building for a reason—they were a mere fraction of what a Marine might experience in a war. Getting through the grueling pit sessions without quitting, staying mentally alert, remembering our infraction...that was the point of the whole exercise. But the stress was too much for someone.

I heard someone a few rows away softly but spitefully spit out the "N" word directed at one of our drill instructors. He had made the mistake of letting his emotions control him, and he had snapped. He wasn't stupid enough to make it loud enough for the drill instructor to hear, but it was loud enough for another recruit to hear, and that recruit quietly told a drill instructor about it. He didn't do it to "rat the guy out," he did it because he knew that word represented a poison that needed to be neutralized if we were going to get through the next ten weeks.

A few seconds later, our Senior Drill Instructor stormed out of the barracks, his face beet red with fury.

"GET IN THE SQUAD BAY AND GET ON-LINE!" He barked.

Within a half a minute, we were all standing in the squad bay at attention, not moving, with our bodies locked at attention near our bunks. The squad bay was a large open barracks room filled with two straight lines of bunk beds. Each bed had green wool blankets with tight hospital corners, crisp white sheets, wooden footlockers...and now sandy, sweaty, and scared recruits.

What's going to happen now? I thought. My eyes slid over to the "special" gap in the line reserved for the offending recruit who was still outside while the rest of us were all on line.

"I have a dream that my four little children will one day live in a nation where they will not be judged by the color of their skin, but by the content of their character."
– Dr. Martin Luther King, Jr.

When defining others, substance is everything. Image is nothing. To most kids (and some adults), though, it's the opposite. Some grow up and realize the truth that what someone looks like doesn't matter, but some never figure it out—they get older and older and become very shallow, sad, and confused adults. These are the people getting facelifts until their ninety or waste their later years wishing they were young again because surface qualities are how they define others and themselves.

Racism is the extreme form of this shallow approach to others, but it includes any superficial and lazy judgment like someone's weight, looks, clothes; the things that children often resort to when defining other people. As leaders, we have to make sure to never set this example and to treat all people the same until their character dictates we treat them differently.

> "The only man who behaved sensibly was my tailor;
> he took my measurements anew every time he saw me,
> while all the rest went on with their old measurements
> and expected them to fit me."
> – George Bernard Shaw

We waited in the squad bay. Only a few of us had any idea of what was going on—the rest were just glad to be on our feet again and out of the sand pit. Eventually, the door in the back of the squad bay opened and the drill instructor walked the offending recruit past all of us to the front of the squad bay in the small open area called the "quarterdeck."

The recruit was commanded to stand at attention like the rest of us. When he passed me on his way to the quarterdeck, I thought he looked like a man walking to the electric chair. The sweat that must have been burning-hot just moments ago was streaming down his pale white temples; I was sure it felt like icy rivers. The rest of us stood at attention, quietly waiting.

Five minutes went by...slowly...and then we heard the squawking of radios and clanging of handcuffs coming from the opposite end of the squad bay. Two Military Policemen walked to the quarterdeck where the recruit was standing. Quickly and without saying a word, they handcuffed him. He was turned around and flanked by both MPs as the three of them walked down the long gauntlet of recruits to the back of the squad bay and out the door. The sandy paths on his cheeks that the sweat had made minutes earlier were now filled with rolling tears.

He wasn't arrested or charged with anything (we were told later), just "recycled" to a platoon that was three weeks behind us in training. It was a far worse punishment than paperwork...and a much more effective lesson. He had to do the second, third, and fourth week of Boot Camp over again as though he had never done any of it at all.

It was an effective way to teach the sorry recruit—and the rest of us—a real lesson that the laziness involved in defining others by anything other than their actions was not going to be tolerated in the

Marine Corps, an organization founded on strong principles and devoted to the formation of character. Just as a family should be.

The senior drill instructor sat us down later that evening in the cool grass outside of the squad bay and explained that only someone's character counts when defining them. He explained that if we ever went to war, we would have to depend on each other and we won't care about someone's color or looks or anything else if our life is in their hands; we will only care about the quality of person they are, their character. With our shaved heads and matching clothes and shared hardship in Boot Camp, we learned to look past cheap images to see the rich substance in each other and we learned that character is what counts. Nothing else matters. And when we begin to see others like this, we can see ourselves like this. We can shed our cheap habit of measuring ourselves by our own looks, job, handicaps, and bank account and start to see clearly what's important about our character. We learn about who we really are, and from there we can improve ourself.

We must teach our children the same principles: we must see each other for who we are, not what we look like. We cannot keep poison in our hearts and be *leaders of character*. Your character is what they will watch to learn to develop theirs. The good self check of your of character includes taking note of both how you act in adversity and how you treat people who cannot hurt you or help you in any way. Do you treat the guy on the corner selling flowers with the same respect you treat your boss? Do you treat waiters with respect? Do you respect those people who are weaker, less popular, less powerful, less wealthy, or more vulnerable than you? Can you tell the truth when the truth will hurt you? Do you stand up for others when ridicule directed at you could be the result? Do you do the right thing when only you will know what you did? The answer to these questions will give you a good idea of the strength of your character. Actions that reflect competence in adversity, truth regardless of consequences, and kindness towards all others signify strong character.

Once you are working on your character solidly, you can begin to teach it not only by example in everyday life, but directly to your kids like our drill instructor did for us that hot summer evening in Parris Island. You can encourage your child ahead of time (during sit-around discussions) to go to the aid of the child who seems lonely or in distress. Your kid can help another child who's being picked on for having things like red hair or big teeth or being fat or whatever trivial quality kids choose to focus on to inflict pain. Your child can stand up for the other child (physical courage and moral courage) despite the risk of his action making him "unpopular." They'll find being a protector is much more rewarding and honorable than being a bully. Some kids do this naturally out of compassion. Others can learn this behavior and practice it. Not all goodness springs from a natural desire—sometimes it must be learned and then practiced. I find I have to practice this often. Our world of "image" is not conducive to celebrating character; we must celebrate it and practice it ourselves.

From our perpetual practice and understanding of the importance of someone's character, strong character in ourselves is strengthened and becomes a ray of love and protection to anyone who comes into contact with it, especially our kids.

"The best index to a person's character is
(a) how he treats people who can't do him any good, and
(b) how he treats people who can't fight back."
– Abigail Van Buren

CHAPTER 6

"Pepper-Spray Yourself"

Developing the Courage and Confidence to Take Action

"Courage is rightly esteemed the first of human qualities
because it is the quality which guarantees all others."
– Winston Churchill

I pried open my aching eyelids with my fingertips and saw that fall's crisp world had turned into a watery, brown blur. The breeze didn't caress my skin with its usual cool touch as it swept by–instead, it turned to sharp gusts and bit into my eyes. It was the first time I could remember resenting a simple breeze. I tried to pry my eyes open wide enough to see clearly, but my eyelids wouldn't move.

Pepper spray tastes bitter and warm, and it's just like eating hot peppers—you're overcome by a wave of painful fire and you feel like you've just been bobbing for french fries that were still in the grease pot.

I had just been sprayed in the eyes with oleoresin capsicum (OC) pepper spray as part of weapons retention training in the first Fleet Marine Force-Wide Marine Corps Martial Arts Program Instructor-

Trainer course in Quantico, Virginia. We were training to become instructors of instructors. Our training exercise that afternoon was to be sprayed solidly with pepper spray and then keep control of our sidearms (using specific techniques) as other students tried to grab the firearms away from us. This scenario would familiarize us with the sensation of getting sprayed so that if we ever *were* sprayed or faced similar pain in real life, we would have a better shot at functioning effectively and maintaining control of the situation.

After the exercise was over, I laid on my back on the cold grass with a sense of relief in my mind and sense of pain everywhere else.

> "You gain strength, courage, and confidence by every experience in which you really stop to look fear in the face. You are able to say to yourself 'I have lived through this horror. I can take the next thing that comes along.' You must do the thing you think you cannot do."
> – Eleanor Roosevelt

Marines are confident because they've been built up mentally with many small successes. These scenarios and successes start in Boot Camp and increase both in difficulty and danger as the recruit progresses and continues into a Marine's career. (The pepper spray drill was not dangerous, but it was difficult.) All of these small tests lead to confidence and competence, qualities that are developed to accompany Marines into war and give them the edge they need to win.

You can build this kind of confidence for yourself whenever you undertake a challenge. Team challenges like sports are especially critical for developing confidence in kids and teenagers; likewise, individual sports like golf and wrestling can also build mental and physical confidence. Contact sports most effectively build the kind of confidence we need to act in the face of physical danger but other challenges that require courage like acting, public speaking, performing, etc (for kids who just don't like sports), can build that confidence too.

Participation in any activity that requires courage—bit by bit, event by event, season by season—is helpful in building courage in us and our children. And we must have this courage! If we do not, our goodness is impotent and worthless because we will not be able to be the leader and take the right action when action is needed.

> "All that is necessary for the triumph of evil
> is that good men do nothing."
> – Edmund Burke

This *courage to act* is a prerequisite to good parenting and honorable adulthood. We must act in the face of fear, danger, criticism or even peer pressure. We must never *do nothing*.

The opposite of not having courage happens when we stand by while we watch someone get beat up. "It's not our responsibility," we say out of fear. We rubber-neck past a new accident on the road rather than getting out and seeing if anyone needs our help. We rely solely only on the police to protect us when we ourselves should have some means of protection for ourselves.

This lack of courage starts when a kid silently stands by as someone bullies someone else…or worse yet, when that child participates in the aggression. Most kids will pick on someone sometime—it's their immature nature—but the sooner they realize it's better to be a *protector,* they'll never go back to the weakness of bullying again. They'll start to find they're earning respect for their courage from others, both the bullies and the bullied. Sadly, though, some people never learn that lesson, and this lack of courage often persists into adulthood.

When your child does make the courageous move to sit with the new kid or to be friends with the lonely kid, the seeds of leadership are planted. That's how children become honorable. That's how they start to become leaders. Little steps in courage are as important as monumental ones.

Honor, Courage, Commitment
– The official Core Values of the Marine Corps

Wow...what a feeling, I thought as I laid there with my pepper-sprayed eyes that I still couldn't pry open. *I can't believe I'm getting paid for this...this is awesome!*

Eventually, the waves of fire subsided and I was able to drive my car back to the barracks. I was glad I'd waited out the pain instead of trying to neutralize it by pouring milk in my eyes the way I'd heard some Marines did. (I'd also heard that some had made the mistake of showering while standing up, which resulted in contaminated water running from their faces to their more sensitive areas...not a good idea.)

Why was I happy after having been pepper-sprayed in the face? Simple: the experience and the confidence and the rush of passing a difficult test had made me feel alive and confident. It reinforced to me the truth that taking those steps of courage always leads to the high of *having more confidence.* It makes you yearn to keep pushing yourself to new limits. You realize there really is no limit for you in life. It's a rush!

It's this rush of confidence that creates more courage and more courage leads to activities that result in more confidence. It's a wonderful ladder to be on and anyone can climb onto it. The Marine Corps doesn't own this ladder, but they sure know how to lead people up it (There aren't many Marine's that lacked self-confidence).

As a parent, even if you have self-confidence and you know you already are setting a strong example, remember the ladder never ends and there's more opportunity for you the more steps of courage you take. If you think you may be lacking the courage to set the *take action* example when it's needed, take that first small step into what you fear to do, whatever it is, and keep doing it rung by rung! All small steps of courage lead to great leaps in confidence.

"The most vital quality a soldier
can possess is self-confidence."
– General George S. Patton

CHAPTER 7

"Stick to Your Guns"

Not Compromising Your Principles

Mission Accomplishment, Troop Welfare
– The objectives of Marine Corps Leadership
(in order of importance)

The hot South Pacific evening wind blew across the runways and through the chain link fence into the complex of tents and Humvees.

"Sir, you *will* stay here until Sergeant Klein gets back!" Sergeant Clarence Davis said sternly to the young liberty-minded second lieutenant. He wasn't a bad officer, just a new one, and Sergeant Davis issued his admonishment with all the tact he could muster. (Though I was also a Sergeant, Sergeant Davis was my senior and the crew chief for the training exercise, so technically I was under his charge.)

Normally, a Sergeant wouldn't talk to an officer that way, but this was no normal Sergeant. Clarence Davis was a former infantry Marine who had fought in Haiti and had more medals on his chest than most Colonels did. He was an excellent leader who understood leadership more clearly than almost anyone I've ever met. "Intense" was the best way to describe him. So were the words "courageous" and "principled." Tact wasn't his strongest trait, but then again, being courageous and principled was probably more important than tact, anyway. Accomplishing the mission and taking care of his fellow Marines were the most important things to him.

"Sir, the training exercise is over, but now it's your job to ensure that Sergeant Klein comes back safely," he explained to the somewhat-startled lieutenant. "You will not go back to the barracks and shower until he lands." The lieutenant wasn't sure what to say—he was teetering between feeling angry that a lower-ranking Sergeant was telling him what was up, and at the same time having a sneaking suspicion that the Sergeant might be on to something.

That particular lieutenant and I had just spent a few days on a mountaintop in central Luzon, the largest island in the Philippines. We'd been conducting forward air-control functions for a joint Filipino/U.S. military exercise called Balakatan. The island—along with the rest of the Philippines—was having some problems with rebel forces in the jungle who were kidnapping and killing people.

When our mountaintop air traffic job was finished, we had to bring our vehicles and radios down the hill for transport back to the base near Manila. We had hired local truck drivers to haul our vehicles back to base as a show of economic support and friendship, but we did have to provide an escort to ensure that they weren't stolen along the way by the rebels, and although it wasn't as dangerous as it sounded, escort duty did require us to be very aware of our surroundings and keep the truck drivers on course.

"Sir, I'll see you in a few hours," I said to the second lieutenant as we went to man our separate vehicles. They were being loaded onto the flat-bed Filipino trucks in preparation for our trip back to the base. From the base, we were to catch a helicopter back to the airfield where our barracks were.

After the convoy had spent a scenic and uneventful three hours traveling through small towns and jungles, we arrived at our destination. Since the lieutenant arrived first, he was instructed to catch the first helicopter back to base, which meant that I was on my own. For whatever reason, after his helicopter left, things really slowed down. The helicopter that I was supposed to ride on had been diverted to do something else, so I just waited with my pack and rifle slung over my shoulders. Waited...waited...waited. Only a couple Filipino Marines and a few others were there; just like I was, they were also waiting for a ride.

It started to get dark. We broke open an MRE or two and the Filipinos had some rice and eggs. We played a few hands of Spades...kept one eye on the jungle around us...and waited some more.

Finally! I thought when the sound of thumping rotors came over the trees. The helicopter landed a few hundred yards away.

"Time for Liberty!" I said aloud. I knew we had that night and the next day off to hit the town of Angeles City for some cold San Miguel Lager and a lot of fun. We all gave a sigh of relief and filed into the dark CH-53 helicopter. It lifted us up and over the jungle and flew us south to Clark Field, toward the lieutenant and Sergeant Clarence Davis and the rest of my fellow Marines.

"A good leader can't get too far ahead of his followers."
– Franklin D. Roosevelt

Sergeant Klein and a Filipino Marine controlling our aircraft.

The following list is the next very important piece of information in this book. It is a list of the Marine Corps Leadership Principles. These eleven leadership principles are referred to as "Marine Corps Leadership Principles," but they will benefit any leader, including parents. Just substitute "children" in place of "subordinates" or "Marines." These will work in conjunction with your study of the fourteen leadership traits from chapter two (and there is also a cut out of these in the back of the book).

Marine Corps Leadership Principles

- **Know yourself and seek self-improvement.**
- **Be technically and tactically proficient.**
- **Develop a sense of responsibility among your subordinates.**
- **Make sound and timely decisions.**
- **Set the example.**
- **Know your Marines and look out for their welfare.**
- **Keep your Marines informed.**
- **Seek responsibility and take responsibility for your actions.**
- **Ensure assigned tasks are understood, supervised, and accomplished.**
- **Train your Marines as a team.**
- **Employ your command in accordance with its capabilities**

Although a whole book could be written describing the value of this list, most principles are self-explanatory and you can decide how to interpret them. This very last principle just means that you shouldn't set your kids up for failure or demand too much of them—i.e., don't give orders that can't be carried out, because after all they are only children and too much failure at a young age can be hard to overcome. We need to be extra mindful of that principle with kids.

> "Moderation in temper is always a virtue;
> but moderation in principle is always a vice."
> – Thomas Paine

I landed at Clark airfield and looked out to see Sergeant Davis and the lieutenant waiting for me beside a Humvee parked on the tarmac. I had no idea of the awkward and tense conversation that had just taken place between them—all I knew was that I was glad to see them. I waved to the pilot, then jogged across the tarmac and shook hands with the lieutenant and Sergeant. Davis. We all climbed into the Humvee.

"Let's get out of here—I need a shower!" I said as we pulled out of the airfield toward the barracks.

After my shower, Sergeant Davis and I caught the open-air, extended-back Jeep taxi (called a Jeepney) into Angeles City to catch up with the other Marines who'd had a head start on the fun. He told me about the conversation with the lieutenant as we jostled along.

He looked right at me with a look of intense conviction. "Always look out for your Marines, Sergeant Klein...always." His words echoed the principle of "know your Marines and look out for their welfare" in the principles listed above. He didn't just talk about principles, he stood for them and he lived them. That's exactly what we all must do, because knowledge of leadership principles and leadership traits is nothing if we don't act on it.

I met his intense Clarence Davis stare. "Thanks," I said. I knew I could learn a lot from someone who lived and breathed our steadfast principles and who always made true on his words with his actions. He truly possessed the leadership trait of *courage* (a Marine could technically be demoted for insubordination by disrespecting an officer). But, he stood for his principles regardless of potential consequences.

Sergeant Davis's sentiments may not have been immediately understood by the flustered second lieutenant, but I took his lesson to heart and carried it with me through the years; since then, I've applied it to my roles as both Sergeant and father. That principle along with the others in the list above, have *always* served me well and I know they'll do the same for you.

Sergeant Klein on left and Sergeant Davis on right;
combat engineers below.

CHAPTER 8

"Sunshine's Story"

Prioritizing
What's Important

"Many men go fishing all their lives without
knowing that it is not fish they are after."
– Henry David Thoreau

I pulled the green bent pins out of the Hummer's tailgate and swung it down so everyone could crawl in. We all seated ourselves on either side of the wooden bench seats in the canvas-covered bed. Once we were settled, I pounded on the side of the bed to signal that we were all set; a second later, diesel exhaust puffed out from above and we rolled down the road and into town. The wind whipped against the rolled-up canvas and blew dust into our eyes.

We were still in the Philippines for the Balakatan exercise. The island of Luzon (where the exercise took place) is the largest one in the Philippines, an island of lush tropical greenness, and warm and friendly people. The capital city of Manila sits on the southern part of the island, nested along the eastern banks of Manila Bay. Across the warm turquoise water of the bay is the port city of

Corregidor. Because it's on the Bataan peninsula, the people living there saw the start of the horrors of the Bataan Death March in 1942.

After World War II was over, the United States established an air base north of Manila and called it Clark Air Force Base. It's in the central part of the island, near Angeles City and close to the road we were following. The United States occupied the base until the fateful day in 1991 when a crack and rumble filled the blue skies and turned them dark. The eruption from nearby Mt. Pinatubo blanketed the air field and much of the island in ten feet of gray volcanic ash.

Though the U.S. had already been contemplating consolidating their Pacific bases, Mt. Pinatubo was the straw that broke the camel's back—the decision was made to pull out of the base and relocate to existing bases in relatively-nearby Okinawa, Japan. Today Clark Air Force Base is an operational Filipino Air Force base that is sometimes used for joint exercises like Balakatan; which is why we were there. Though the infrastructure of buildings/hangers/base housing is still there, just a hint of the past hustle and bustle remains, and the base is more depressed than it is vibrant. It is, however, still operational, and it's staffed by some wonderful people.

We rolled to a dusty stop, and the "A" driver (the one in the shotgun seat) unbolted and swung down the heavy green tailgate. I knew it was only a matter of minutes before we'd be surrounded by short and savvy salesman, i.e., the neighborhood kids who try to make some money selling little trinkets and souvenirs during a military exercise. As soon the tailgate swung open, I saw a sea of faces, all upturned, urgent, and primed for their sales pitch. It seemed like today's main product was wooden flutes: the asking price was "one dollar, U.S." Though I am not an avid flute player, the little instruments did sound cheerful, and I thought a flute would make a great souvenir.

I bought one flute and then tried without success to convince myself I only needed the one I had already purchased when a boy showed me that two could be played at once. As the choppy sea of wiggles, hands, and flutes started to calm, I noticed a young girl— about eight years old—standing in the back amongst all the boys.

She didn't just stand out because she was a young girl among the boys; there was something else about her that caught my eye. She looked at me, gave me a great smile, and said "Hi!"

The loud calls of "meester, meester, look!" almost completely subsided as the boys pegged me for a lost cause—they shifted their efforts toward the other Marines I'd come with who were keeping an eye on the Hummer.

The little girl was still standing there, about ten feet away. She called out to me. "What's your name?

"Max" I said. "What's yours?"

"I'm Sunshine," she said quickly.

"And how are you doing today, Sunshine?"

"Just fine, Max," she said. Her English was pretty solid (far better than my Spanish) and I could understand what she was saying without any difficulty.

I walked over to where she was standing and chatted with her a little about the Humvee. As I practiced some of my Spanish with her, she laughed at my fumbled wording and choppy accent, but she helped me along a little at the same time.

After a few minutes, she asked to see the camouflage hat I was wearing. I promptly handed it over to her.

She tucked her dark hair back behind her ears, put the hat on very carefully, and looked at me for approval, smiling brightly all the while.

"Well, that looks good on you!" I said.

She turned and began pointing at different streets and alleys and buildings around her town, telling me all the places where I could get this drink and that food and anything else she thought I might need. She pointed to the street where her parents lived and where her cousins lived. She just seemed happy to be talking to someone and happy to be alive.

As she talked, I started to wonder how she could be so happy when she was out on the street all day selling flutes and hoping for a dollar or two, waiting for hours and hours for the next group of Marines to pile out of their trucks. Huh, I thought, come to think of it, she hadn't made a strong effort—if any—to sell me a flute. She seemed more interested in talking and in my fumbling efforts at communication than at making a sale; she seemed, in fact, to be having a grand time correcting me.

Soon, the other guys came out of the store (six cases of San Miguel Lager for $27!) and threw the cases in the back of the Humvee. We hopped back in. Before leaving, though, I bought a flute from the girl and told her I'd stop by again soon, even though I didn't think I would really see her again as there were so many children in town. But it was nice to see that someone who had so little could be so happy—she just proved to me there is no relationship between material goods and happiness.

"The world has to learn that the actual pleasure derived from material things is of rather low quality on the whole and less even in quantity than it looks to those who have not tried it."
– Oliver Wendell Holmes

As the Beatles said, "Money can't buy me love." The wisest human beings know that while material things may make life a little easier, in the grand scheme of life, things are not very important _other than what they can do to facilitate our relationships with our families and friends_.

But, we must never let our house or car or job or clothes define us—only our character matters.

Many of us strive all our lives for more *things*, but if things are our top priority, we will be sad, unsatisfied people even when we have all the stuff we want. "Having it all" won't change anything—ask the millionaires who are on drugs or depressed. There is no correlation to true happiness in life and the amount of stuff we acquire. It's far better to put others first, and put relationships first (with God, family, friends, others). That's when all other *stuff* will fall into place.

Working hard for things and having money is *not* a bad thing at all, it's the American Dream! You just need to keep priorities in line to make sure it's a good dream.

> "All our lives we sweat and save,
> building for a shallow grave."
> – Jim Morrison

For the next few days of the Balakatan exercise, I was on a hilltop over Crow Valley, a bit farther inland. Mt. Pinatubo loomed large on the southern horizon; ahead of us, a large valley of dust pinned in by two mountains absorbed multiple training bombs and 50 caliber rounds from our joint aircrafts. For a few days, a Filipino pilot and I stood in a makeshift watch tower with an HF radio to keep each one of our aircraft away from its sister planes.

I caught a CH-53 cargo helicopter back to the field and was ready for some time off. After a quick shower and chow at our "villa," we stopped off in the small town again for some more bargain-value supplies. Once again, we found ourselves driving down the street in search of a parking spot big enough for the Hummer. (In a land of Datsuns, this is not an easy task.)

By the time we found a spot, the flute pushers had already spotted us. I'd almost forgotten about having met Sunshine—she'd become another flash of memory in the action-packed days of the Balakatan exercise. But as the e-brake clicked into place, from the back of the throng of enterprising young salesmen, I saw a frantic wave and heard a cheerful and friendly "Max, Max, hi, Max!"

It's Sunshine! I thought. *How did she remember me and pick me out right away out of all these guys???*

We talked a bit more and we told each other what we each had been doing over the last couple of days. By that point, I just plain liked the kid—she was full of spirit and happiness and was bubbling with conversation. She told me about what she was learning in school and about her family who loved her so much. I decided that was why she was so happy. She asked to wear my hat again while we chatted. I happily obliged, of course.

That day was the last day of the exercise; we were about to leave for good. Before we did, I decided to give Sunshine the camo hat since she liked to wear it so much. She smiled from ear to ear, then gave me a big hug and said, "Thank you, Max!"

I wanted to thank her as well. The spirit of her pure, solid heart shone through her eyes and her wide smile, and her actions sparked warmth in those around her. According to our conventional standards of living, she should have been unhappy in her daily struggle to earn money—many of her efforts, I'm sure, ended in vain—but instead she was a happy girl. She wasn't happy because she didn't know any better (about what it's like to have material things), she was happy because she knew better than most what is important. It's much easier to see these things when material things don't consume all your vision! As the Bible says, "Again I tell you, it is easier for a camel to go through the eye of a needle than for a rich man to enter the kingdom of God" (Matthew 19:24).

It isn't that we can't be rich and good, but there's no doubt that it's easier to see the truth of life when worldly (and ultimately cheap) *stuff* is not your priority. Sunshine valued relationships and people, not "stuff." She appreciated a gift, she appreciated a conversation, and she seemed to absolutely love being alive. She seemed to know how fortunate she was to have a family who loved her very much. Her example reminded me that real happiness and contentment is not found in *more stuff*. Many of us, you and I, were not well off when we were younger, but did we care? Did our parents make us feel like it mattered? Probably not. Many of us weren't rich but we were happy. Why not take that example into our adulthood?

Having more for more's sake is a quest without reward. Having more to enable you to have more quality time with your family is why we should want "more." Put down that list of material priorities and realize that richness already exists right in front of you—any stuff on top of it is just icing on the cake. This way, your happiness is not dependent on your bank account, and should you *lose it all* someday, you'll still be content.

For me, a dusty old camouflage hat given to a little girl in an obscure town across the sea and a few colorful flutes are symbols of a brief but genuine friendship in a far-off land. It was a chance encounter turned to gold. Not only is my memory of that encounter the best souvenir I have from the Philippines, it was an experience that made little Sunshine and I very rich people.

> "That man is the richest whose
> pleasures are the cheapest."
> – Henry David Thoreau

Little Sunshine (center) reminds us what's important.
Sergeant Clarence Davis is on the right.

"But seek ye first His Kingdom and His righteousness,
and all these things will be given to you as well."
– Matthew 6:33 N.I.V.

CHAPTER 9

"History's Handbook"

Choosing Your Role Models

"Few will have the greatness to bend history itself; but each of us can work to change a small portion of events, and in the total of all those acts will be written the history of this generation."
– Robert F. Kennedy

The warm, sweet air of Tinian Island filled its lush jungles and open emerald fields, providing a fragrant backdrop to the surrounding ocean's alternating deep and light hues. The sky above was a bright indigo. Tinian—a volcanic speck of an island in the southwest Pacific Ocean—had been once been used as a base for U.S. bombers who'd then flown on to drop the atomic bombs that had ended World War II.

During the war, an airfield had been built by the Seabees (or Navy combat engineers) for the impending attacks on mainland Japan. Just across where the Philippine Sea and the Pacific Ocean meet was the island of Saipan; it was only a few miles away and was visible from Tinian's shores. The more well-known island of Guam was a little bit farther than that.

Tinian's old runways were cracked from years of neglect; clumps of grass sprouted out everywhere.

Unused runways—with their solid, open, and grown-over expanses—harbor a strong sense of history. It's easy to feel a connection to the past when there's no difference in how things looked then to how they look now. On Tinian's runways, only a few cracks and some wandering grass bore testament to the passage of time.

"All right, stop! School circle around me!" I yelled as the last exercise of the conditioning drill was completed. I was conducting a grey belt Marine Corps martial arts course for about 20 Marines during our month-long exercise on the island. The students and I had found an open field near the runways that was suitable for training. They had just finished about a half hour of grueling conditioning, including grappling, fighting, and many other exercises to keep the body fatigued so that technique was the only option left at their disposal. That kind of practice is one of the best ways to emphasize focusing on technique when teaching martial arts: when a student is too tired to brute-force his way through the motions, he must rely on the correct technique, just like when we as parents are run down by life we still have strong principle to rely on to keep us effective. The other reason for whipping them into a state of extreme physical fatigue was to simulate the fatigue of a war-time scenario that they may encounter someday.

Students are also receptive to character studies when they go from a state of fatigue to one of relief.

Part of the Marine Corps Martial Arts program is character training, including taking in and understanding events associated with not only Marine Corps history, but history of other "warrior cultures" as well. This character training is presented in talks by the instructor called "tie-ins" because they subject matter *ties in* to the physical character training that the student is experiencing. I knew, as the instructor, that actually physically being on the same land that the lesson would be about would have a great affect on the hearts of the students.

With sweat running down their faces, their hearts still hammering in their chests, their shirts soaked, and a glow of pride in their eyes, the students sat down around me in a circle and prepared to learn from the "character" tie-in portion of that day's class. One by one, they loosened their flak vests, took off their helmets, and let the cool breeze push through their wet green T-shirts.

"Close your eyes and listen," I began, then read on.

Private First Class Robert L. Wilson

Rank and organization: Private First Class, U.S. Marine Corps. Born: 24 May 1921, Centralia, Ill. Accredited to: Illinois. Citation For conspicuous gallantry and intrepidity at the risk of his life above and beyond the call of duty while serving with the 2d Battalion, 6th Marines, 2d Marine Division, during action against enemy Japanese forces at Tinian

Island, Marianas Group, on 4 August 1944. As 1 of a group of Marines advancing through heavy under-brush to neutralize isolated points of resistance, Pfc. Wilson daringly preceded his companions toward a pile of rocks where Japanese troops were supposed to be hiding. Fully aware of the danger involved, he was moving forward while the remainder of the squad, armed with automatic rifles, closed together in the rear when an enemy grenade landed in the midst of the group. Quick to act, Pfc. Wilson cried a warning to the men and unhesitatingly threw him-self on the grenade, heroically sacrificing his own life that the others might live and fulfill their mis-sion. His exceptional valor, his courageous loyalty and unwavering devotion to duty in the face of grave peril reflect the highest credit upon Pfc. Wil-son and the U.S. Naval Service. He gallantly gave his life for his country. (Source: www.history.army.mil)

"Keep your eyes closed," I said softly. "This is where it happened—a Medal of Honor was won on the ground you're sweating on now." I paused and let a brief silence settle over us.

"Open your eyes," I finally continued, "and look around. THIS is where it happened, in these trees and near this field. We're not far from these actions—only time separates us. It could be those same rocks that the enemy hid in. Robert Wilson was a Private First Class and 23 years old, not much different than any of you."

"Marines, we have a legacy to uphold. We are doing all this training for a reason. You are learning to persevere through pain for a reason. You are resting now because you worked together as a team. PFC

Wilson loved his brothers and lived to protect them and was willing to die to do that. He had courage to 'precede' the advance to the enemy position. He found a way to keep the others alive. Brave. Selfless. Courageous. His blood lives on in us...

"Marines like him are the ones we strive to emulate, so when you are on the edge of giving up on doing the right thing because things are getting tough—whether it's training harder, telling the truth, or putting yourself in harm's way to protect others—remember PFC Wilson. You are like him. You have to remember to find a way to do the right thing, to stay honorable, to be courageous, and to be committed to his legacy and your fellow Marines. We Marines are tenacious and we never, ever give up!"

"I'm proud of you because I know all of us ARE upholding Private Wilson's legacy."

"The supreme purpose of history is a better world."
– Herbert Hoover

Sergeant Klein teaching knife-fighting techniques on Tinian.

One of the most important things we can show your children is that we can learn about ourselves by looking at history and choosing role models from it. It's a lot easier to know where you are when you know where you have *been* as a country, as a culture, as a family. History can also make everyday things fascinating. During our free time in Tinian, we hiked through the woods and saw pill boxes and fortifications that had become overgrown with weeds. Those structures probably hadn't been seen since the Japanese had been driven from them during the invasion.

With a little imagination and effort, we can really understand history. We can feel history—see it and touch it. Reading and learning about history is to life what goggles are to divers: windows to a deep world of wonder, power, and mystery and a place where you can meet and learn from the most incredible human beings that ever walked the earth.

History is all around us and the role models we can pick from it are ripe for the picking. Some of us will never be interested in it, but for those of us lucky enough to find its fascination, the world becomes a much richer place.

Diversify your influences to honor your children.

Later that week, as the class was training in some ground-fighting techniques, I glanced up the dusty road to see a group of older men walking slowly down it, looking around as they made their way towards us. The young uniformed Marine who was preceding them led them down to where we were training.

"Keep practicing that technique—I'll be back," I said to the class before I turned and walked up to the young Marine.

"Sergeant, I have some Marines here with me who fought on Tinian," he told me. "They'd like to observe some of your class."

I'd had no idea these Marines were coming for a visit. No doubt it had been a trip organized by Public Affairs in conjunction with our exercise. Some of these guys may have even known PFC Wilson, I realized. What a great thing to have them here!

I looked each one of them in the eyes and shook their hands. "It's an honor to meet you Marines," I said. "I'll continue with the training, but please stop me if you have any questions." I shifted my attention back to the class.

I told the class who we had in the audience and their intensity heightened. "Back to back!" "Grapple!" Grass flew, teeth gnashed, and grunts emanated from the swarm of camouflage, soil, and sweat.

"Switch partners! "Back to back!" "Grapple!" I barked as the gritty volcanic soil flew.

A few minutes into drills, I glanced over and noticed that the older Marines were definitely not bored.

Judging from the way their eyes were lighting up, in fact, I thought one or two of them might go at it themselves.

"Ten Marines here, ten Marines there!" I yelled as the two groups of students went to opposite sides of the open field.

"We're going to end this class with bayonet techniques!" I figured it'd be a good way to end class on a motivational note. To build *esprit de corps* and stoke the fire for next class, I'd line up two opposing "armies" across the field and have them both advance on each other while performing bayonet techniques and doing all they could to strike fear into the opposing side...I told them that whichever side made me crap my pants in fear would win.

"Fight!" I yelled.

All hell broke loose. Screams roared and bayonets slashed as the armies closed in. When they met on the field, they "killed" each other with a viscous tenacity that always ended in hearty laughs and high spirits.

"All right, Marines," I finally instructed, "get your gear! Get in formation and we'll run back to the airfield."

Later that week, I heard from an officer that the old Marines who had fought on Tinian had been very impressed with the class. Visiting Tinian had been an emotional and fascinating experience for them in general, they'd said, and they'd also said that the best part was seeing all the young Marines training. All of us—the pilots, the grunts, the radio operators, and all the others—training hard with *esprit de corps*

The old Marines were happy. I think they knew we still had their fighting, can-do spirit. Because the Marine Corps values its history, we honor these men—these excellent role models—by emulating them, and the world is better for it. And the world will be better also when you choose the right people to emulate.

Depending on the role models you choose from world history or your history, your efforts will also make the world a better place. We must wisely chose our role models from history and from our lives, whether those role models be our parents, grandparents, friends, someone we read about... we need keep the best and drop the rest (even if "the rest" is our own parent's influence). Find as many sources of inspiration as you like—history is rich with courageous role models. The folly and heartache in parenthood lies in choosing only one. So if you realize you came from a shaky tree, do the honorable thing and make sure your apple falls far from it so a new tree of beauty can begin to grow.

Private First Class Robert L. Wilson, USMC

CHAPTER 10

"The Action Example"

The Importance of Practicing What You Preach

"It is not the critic who counts; not the man who points out how the strong man stumbles, or where the doer of deeds could have done them better. The credit belongs to the man who is actually in the arena, whose face is marred by dust and sweat and blood; who strives valiantly; who errs, who comes short again and again, because there is no effort without error and shortcoming; but who does actually strive to do the deeds; who knows great enthusiasms, the great devotions; who spends himself in a worthy cause; who at the best knows in the end the triumph of high achievement, and who at the worst, if he fails, at least fails while daring greatly, so that his place shall never be with those cold and timid souls who knew neither victory nor defeat."
- Teddy Roosevelt

Marines literally *in the arena* passing time as usual.

"Are you ready to kick my butt?" I shouted to the class. They had no doubt felt the urge to do so during the previous weeks of numerous training exercises.

"Hoo-rah, Sergeant!" came the confirmation that they did. The grassy hills of Camp Pendleton in southern California surrounded the matted-down field; above the field, the hot dry sun beat down on the heads of the students. I was teaching a reserve unit during their two-week annual summer training.

As a martial arts instructor I had to show the students that I knew what I was teaching. Your credibility as a leader skyrockets when you can actually *do* what you're teaching and perform at the level you're demanding of your students.

"All right—one at a time for one minute each," I said to the 15 or so students who stood ready for the challenge. It was a kind of tradition to take on the whole class one by one at the end of a Gray belt course.

The first student and I sat back to back, and the Marine with the timer said, "Go!" We wrestled using the ground fighting and submission techniques they had been learning—arm bars and chokes from the mount or guard position. I had the edge on the first few Marines, but I also let them work any strong techniques they'd started to develop.

As the minutes passed, having to wrestle a fresh Marine every other minute began to wear me down, which meant I had to rely on technique. (That was part of the lesson, too.) Pretty soon, I was getting so fatigued that the technique was tough—even on a fresh student—and they began to get the edge. I didn't mind that at all—if your student kicks your butt or your child exceeds you in competence in anything in life, it just means you're a good teacher.

One minute of *doing* is worth more than an hour of *saying*.

We need to be able to set an example by living it or *practice what we preach*. No one respects someone who can only talk about something instead of doing it; likewise, a child won't respect a parent who says, "Do as I say, not as I do". We have to *be* the person we're telling our kids to be, not perfect by any means, but honorable and principled.

You've got to remember that your kid probably wants to be just like you and may want to be just like you for the rest of his life: you're their first hero and probably their last. Bearing that weight—and seizing that opportunity—is one of the most important tasks as a parent. *Do it right* and your kids will follow. Of course, we will fail at things along the way, but failure is just a new opportunity to improve. Remember, failing at a noble task is always greater than knowing "neither victory nor defeat"

"Example is not the main thing
in influencing others, it is the only thing."
– Albert Schweitzer

Scrawny Student #15 had me in a choke; stars began to whirl at the edges of my vision.

"Tap!" I mouthed silently—there was no air to make a sound. My free hand tapped on the student, he released me, and sweet air came rushing back into my lungs.

"Great job!" I said. "That's how you sink a choke!"

The morale of the class was high: they'd each gone one-on-one with the instructor and the techniques I was teaching them really did work. They also were more receptive and respectful students after that experience because they knew I could "walk the walk" with them and not just "talk the talk."

All military leadership skills are based on the principle that you can do, have done, and are willing to do what you are asking others to do; in other words it's all about *setting the example*. It is exactly the same with raising a child. A child will eventually have the courage to protect someone from a bully if they see you do it. Someday, they will speak up in spite of ridicule if they see you do it. They will do the right thing when no one else is looking when they see you do it when only they are around. And they will also have fun and laugh with life whenever possible if you do the same. A

child will lead their own children someday by watching
how you do it today.

> "As I get older, I pay less attention to what men say.
> I just watch what they *do*."
> – Andrew Carnegie

CHAPTER 11

"Immersion"

The Importance of True Education

"Knowing others is intelligence; knowing yourself is true wisdom. Mastering others is strength; mastering yourself is true power."
– Lao-Tzu

"I want all the Marines trained and refreshed in small unit leadership, patrolling, radio work, and land navigation over a three-day period."

"Understood, Gunny, Where do we get the...?"

"Just do it—you'll never learn if you don't do it!" he said "Get 'em all up there [to the jungle in northern Okinawa] and train them!"

"Roger that, Gunny," I knew that was then end of the conversation. I wanted to ask where you would go about getting permits for transporting ammunition through Japan. Where and what are the Marines eating? What vehicles and radios do we need? Things like that.

As I left the office he said, "I want it all taken care of by the end of next week."

Frustration was knocking at my door at the thought that I was getting zero help coordinating this training exercise. My Company Gunnery Sergeant had just put me in charge of coordinating the scheduling, embarkation, training, billeting (sleeping arrangements), and safety for roughly 120 Marines for a three-day training exercise.

Since I had just been promoted to Sergeant a week prior, this kind of thing was my job now and I had better get used to it. The problem was, I had never really done any of that kind of logistics planning before in Japan, as I had just arrived there two weeks prior. I barely even knew the people in my unit whom I thought might know answers to these questions...but none of that mattered, because I had to *find a way* to manage the situation and get it done. It was an order and I was a Sergeant. That didn't keep me from cursing under my breath, though.

> "Parents can only give good advice or put
> them on the right paths—but the final forming
> of a person's character lies in their own hands."
> – Anne Frank

Education, true education, is not the regurgitation of facts on a test—it's much more than that. Education is not supposed to be the binge-and-purge process it has become in many schools. It is the absorption of knowledge through your own experience (or through another's experiences by listening to them or reading about them). You can become educated without sitting in a classroom or picking up a book, or you can become educated in a classroom without gaining experiences of your own but the only way to become fully educated, though, is to combine both paths: learn for yourself and learn from others. Immerse yourself in the subject. Only fools think they can't learn from others, and only fools think they can learn it all from a book or a lecture. When you absorb the hybrid vigor of experience and theory, you begin to really learn about yourself, which is the top goal of a well-rounded education.

The first two words in the list of Marine Corps Leadership Principles are "know yourself." Those words hold true for everyone, not just Marines: to be a leader and to be truly educated, you need to find a way to know yourself.

What does that even mean? Hundreds of years ago, Socrates said, "know thyself." To this day, many consider his instruction to be the most fundamental and the most important nugget of philosophical insight to have ever stemmed from a human mind. But why is this command so important?

To answer that, imagine that you're alone in your bed. Look around. The setting sun is beating through a yellow faded curtain; it's falling on the night stand in such a way that all the dust particles have shadows. You lift your aged hand up into a ray of light and see the wrinkled skin and bluish veins outlined by the golden dusk. You've gotten old, you realize.

You want to talk to someone, but no one is there. Your family is gone—most have died peacefully of old age, and the grandkids live far away. They love you, but they have busy lives and can't be with you all the time. You close your eyes and think back...perhaps about your wife and how she looked when she smiled in the mid-day sun, her kind eyes sending pulses of love into yours. But she's gone now, too; only the memories are there.

A tear trickles down your cheek....but then a smile stretches across your lips and catches the falling tear, and you welcome the memories. You thank God that you have them and that you had a chance to make them. Life is beautiful even when it's sad.

You hear some rustling in the hallway. A few seconds later, a nurse walks in. Instead of dismissing her quickly, you ask her to hold your hand. She does, although only for a moment—she has too much to do already. You close your eyes again and hum a song, then murmur, "Thank you, God, for giving me this life—there's still more I can do, I know." You open your eyes and smile at the nurse who usually talks over-loudly to you and calls you "mister."

"Thank you for being a good nurse and a kind person," you say, quietly and sincerely.

She stops moving around, looks into your eyes, and says, "You're welcome."

The small seed of the goodness of your words has been planted lightly in her heart—when she leaves the room, she feels a little bit better than she did when she walked in, and that's because of you. And because she feels a little bit better, her partner and her kids will, too.

Later that night, you wake up and feel a pain. You look at the warm glow of the night light; your eyes get wide and your breathing becomes shallow...and then you die with a little smile on your face.

But your soul's not dead, and your soul is what matters. Your life still pulses in waves through people's lives, small ripples that will reach far across humanity and into the future. The world is different because you lived.

The question is, is the world you left behind better or worse because of your actions? If you make an effort to "know yourself" during your life, and you make an effort to gain a true education by immersing yourself in your own experiences and learning from the experiences of others, then you'll always improve yourself and everything around you bit by bit. You will begin to reach your personal potential. You can only lead yourself so far toward your own potential if you don't really understand who you are. Our job as leaders and parents is to approach our potential to set the best example possible. When we do, we will be content with the life we lived because we will have shed wasteful traits and adopt productive ones. We will have appreciated the real things life has to offer, and would have followed our dreams, and will have really truly loved people because we loved ourselves first. And we'll be planting those seeds of goodness in others until the day we die. And then we'll be okay with dying—we'll only be sad for the ones who will be sad once we're gone, not for our own sake. We will know for sure that we've led our family to more honorable lives and that our influence has taken root and will be growing after we have gone. It's like the peace of mind of a builder who knows he's built something beautiful and future generations will appreciate his work even though he's gone. We'll have done our jobs well and be content with moving on. There's nothing left to do now except let our souls keep loving people into eternity. No bitterness will haunt us in our final years. Knowing yourself, this highest goal of real education, makes living much richer, our influence much broader, and dying not so bad. That's why it's important to "know thyself".

Knowing what makes you do what you do and why you are who you are is the first step in becoming truly educated and having a great impact upon others, especially your children. That's why it's important to question yourself. Question everything you think you know, even if doing so is uncomfortable. Constantly refresh your knowledge by taking inventory of your opinions and habits. Are they worth keeping? Should you get a few new ones and chuck some old ones? Are the carrying costs of keeping them just too high? Learn from others, read about others, and learn for yourself...all the time...every day until you die. Education shouldn't only be thought of as a young person's opportunity. Think!

One powerful tool we can use to get to know ourselves from the "learning from others" pillar of education is reading the right books. Reading is a love we must instill in our kids, and we can do that by reading to them. This is so critical that it cannot be overstated. It's hard to improve if we're only looking at life from our own perspective—we need the perspectives of others, and if we can't meet those people and hear their views, we can read what they say.

We must read to and with our children because reading will help them develop the imagination it takes to be true learners and to start developing their own broader views on the world. Books will also give them a firm understanding of language and get them interested in life in general. Encouraging our children to love books by reading to them is one of the greatest gifts a parent can ever give a child. Very few people who have ever changed the world—even a little bit—were *not* influenced somehow by reading.

I'll use myself as an example (not that I have changed the world, but that my life has been positively influenced by reading and I feel I am on the road to *knowing myself*): I joined the Marine Corps partially because I read *Marine*, a story about Chesty Puller, and I wanted to be like him and be part of the Marine Corps...I'm so glad I did that. I wrote this book because I read *The Purpose Driven Life* by Rick Warren and realized I should do what I'm passionate about, whether I'm financially successful or not, I should do what I love to do and things will come from it...not just trudge through life settling for less. I understand my wife and son and friends and clients much better now that I've read *How to Win Friends and Influence People* by Dale Carnegie, an absolutely amazing book about understanding human beings in general. I am fascinated with all of the different cultures of people and places in history from reading novels by James Michener. Hundreds and thousands of books speak to us and become part of who we are. The books listed above are just my list; your list will be completely different. Books show us that human beings throughout history have had the same worries and have made the same mistakes that we have. Books tie us together and help us learn about ourselves, and books like the one's listed above can help keep us from making the same mistakes that were made in history. Some truths are universal and if we don't have to spend years learning them completely from scratch through trial and error, our personal development can progress much more quickly. If only each generation built on the knowledge of the prior generation rather than having to "learn for themselves", what a wiser and more peaceful world we would have. The world may never do this, but you surely can.

Even if we don't read much ourselves, as parents, we need to do our part to give our kids opportunities to broaden their minds through books; we need to show our kids that reading and education is much, much more than book reports and fact regurgitation.

Experience has been a big part of my personal education, too. I felt that I could write a book because the Marine Corps proved to me that I can do anything I want to do within reason if I just work hard enough and stick to it. I embrace life every day because I saw sad people and dead people in other parts of the world, and knowing how wonderful being alive and free is. I wish everyone could see poverty and death up close; it makes people more thankful for the little things we have that we sometimes take for granted like food, doctors, a roof, our families, and each other.

The best way to jump-start your own real education is to start doing things that jolt you out of ruts— those habits of comfort, self-doubt, and stagnant opinions. Transform yourself through challenging experiences, read to gain a deeper understanding of life, talk to others and find out how they've handled life's ups and downs. See and do as much as you can and think about where you've been and where you're going. You will begin to feel awakened—that you're consciously living your life—rather than feeling like you're just riding out humanity's confusing storm. Just try to learn a little something new every day! That effort alone puts you ahead of the pack.

> "I do not think much of a man who is
> not wiser today than he was yesterday."
> – Abraham Lincoln.

Good education happens when curiosity or necessity drives a receptive mind into new situations. *Great* education happens when you seek out wisdom on a subject before you have to find things out the hard way. As Socrates knew, the *best* education results in *knowing yourself.*

> "I have never let my schooling interfere with my education."
> – Mark Twain

Thank God for mistakes and failures—they're the best shortcuts to knowledge and success. After a week of scrambling, scheduling, and coordinating—and making mistakes—I was ready. The exercise happened; the mission was a success. The training went well and everyone was safe.

As I was riding south in a Humvee in the convoy back down the highways of Okinawa, I realized why Gunny hadn't been more helpful—or rather, why he hadn't been more helpful in the way I wanted him to be. As the Chinese proverb says, give a man a fish and you feed him for a day, but teach a man to fish and you feed him for a lifetime. Gunny was teaching me to fish. He taught me ten times the amount I would have learned if he had "helped" me. Plus I was now a Marine Corps Sergeant; I better be able to get things done. I didn't just hold the rank of Sergeant, I had to be able to do what a Sergeant needed to do. He was helping me by *not* doing my tasks for me—instead, he had challenged me to truly learn. I felt like a Sergeant after that...Gunny knew exactly what he was doing.

I took away two very valuable lessons from my experience with Gunny: first, never listen to those voices in your head that tell you that you can't find a way to get something done—they're almost always lying—and second, the more you immerse yourself in new experiences—whether you fail or succeed—the more you'll learn. Even failure can lead to success; in fact, it's often the prerequisite.

Gunny's lesson took me one more step closer to knowing myself and one more step closer to attaining my true education...and one more smile I'll have on my lips someday as I'm lying in my bed waiting for the kind nurse to see me... content with my life.

> "Experience: that most brutal of teachers.
> But you learn—my God, do you learn."
> – C.S. Lewis

CHAPTER 12

"Toys and Taps"

The Value of Serving Others

"Only a life lived for others is a life worthwhile."
– Einstein

A sloping green hill dotted with tombstones sprawled out before us, crowned by a bright blue sky.

"READY, AIM, FIRE...(CRACK)...FIRE... (CRACK)...FIRE...(CRACK)!" The shots echoed through the hills.

"Pooooorrrt—ARMS!" The rifles fell across our chests. "Pre-sennnnnt ARMS!" the rifles were extended vertically out in front of our bodies for the salute.

The bugler's first note of "Taps" pierced the hearts and wet the eyes of the family we were there to serve. The last note echoed, then faded slowly into silence.

We stacked the weapons and marched into place near the casket. "Fold the flag!" came the gruff command from our squad leader, a retired soldier in his mid-60s who stood on the far side of a freshly-dug grave. Heat swirled through the green funeral home tent and around the mourners dressed in black. Ever since I'd gotten out of the Marine Corps, my new job has allowed me to be in the Cumberland County Honor Guard in Pennsylvania. Most of the other guys are retired veterans from the eras of WWII, Korea, Vietnam, peacetime, and now Iraq.

All eyes were on us, and all hearts were with the man in the casket. We didn't know in which war or at what time he'd served, but none of that mattered. We were here now as a service to him and his family and as a thanks for his service. The family huddled silently around the casket.

Inch by inch, we passed the flag to the two veterans who were folding it, dropping our hands to our sides as it passed by. I stared off into the distance again while the flag was being prepared.

I clasped my white-gloved hands on the top and bottom of the triangle-shaped flag and passed it toward the squad leader, who in turn held it with respect as he lifted a round of expended brass.

"Duty," he said as the hot wind whipped the tent flaps and the leaves rustled above.

"Honor," he said as he tucked the second round into the folds of the flag.

"Country." His firm voice carried the words directly to the hearts of the family around him. Everyone—young and old, seasoned veterans and those who have never served in the military—feels something primal and special about those words, something larger than life that transcends time and is bigger than our daily routines and worries.

"Cen---ter----FACE! Prepare for Final Salute! Presennnnnt ARMS!"

"Ma'am, this flag is presented on behalf of a grateful nation as an expression of appreciation for honorable and faithful service rendered by your loved one."

"Thank you," she whispered with a quiet and trembling voice.

> "The only ones among you who will be really happy are those who will have sought and found how to serve."
> – Albert Schweitzer

The military is often called "the service" because those who are part of it are serving the country and the people in it with their time and sacrifice and courage. But performing a "service" applies to anyone who's giving their time or effort to another person; anyone who's sacrificing anything for the sake of others.

A few weeks before Christmas, I met a mother with two young children at a "Toys for Tots" distribution point who—alongside her children—was handing out toys with us. (Toys for Tots collects toys throughout the year and hands them out to disadvantaged families at Christmastime so the kids of these families can get nice gifts.) I asked her if she had been in the

Marine Corps or if her husband had, and she said no,
neither one of them had.

"How did you get involved in Toys for Tots?" I
asked.

"I was looking for a way to show my kids the importance of serving others," she said, "and I thought
this sounded like fun."

I was fascinated with this woman's good-hearted
nature and her desire to teach her children the value
of helping others. What better gift can you give them
at such a young age than to show them the joy that
goes with giving and the humility involved with
serving someone who has less than you in terms of
material things? Since the nature of children is to
think about themselves first, volunteering or any kind
of activity that benefits others and has with no financial or material rewards is a great way to show your
children the true meaning of service and the moral
rewards associated with it.

You won't have to look far—people always need
help. It's easier than you think to volunteer. Call a
soup kitchen or a church or the Toys for Tots coordina-
tor in the area, volunteer for the honor guard or a
coaching position, or go to a graveyard and ask to put
flags on the graves at Memorial Day with your kid.
Your local municipality, VFW, or church can often
point you in the right direction.

The joy that's gained from performing even a small bit of service far outweighs any perceived inconvenience of "not having enough time" or "not knowing how to get involved." Even volunteering once a year for a few hours is a valuable experience, and your children will carry the lessons they gain in those few hours with them forever.

We don't always get recognition from others, but it's not necessary unless your actions are just for show—if we get recognition, that's just icing on the cake. We should, though, give kids some small recognition when they help out as it will encourage them to participate again. But, the real reward they will eventually find out, goes much deeper than recognition.

Spending time with your children, teaching them what "service" is, showing them the meaning in doing things for others and not just for yourself...can be one of the best things we can do as parents. And plus, this service to others acts as education for us and aids us in our continuing quest to "know ourselves."

"No man has ever risen to the real stature of
spiritual manhood until he has found that it is finer
to serve somebody else than it is to serve himself."
– Woodrow Wilson

"You cannot help someone get up a hill without
getting closer to the top yourself."
– General H. Norman Schwarzkopf

The old veteran's family slowly trickled back to their cars and back to their lives with a special place in their heart for the service that day. An older lady approached and said how thankful she was that we were here. She had tears in her eyes and sincerity in her voice.

"You're welcome, ma'am," is all I said in reply. Participating in a funeral service is a special experience—even if people don't say "thank you"—because you know how much your presence is appreciated. It is such an important service to the families of the fallen.

We gathered the expended brass rounds and gave some of it to the family, packed up the van, and then headed our separate ways back to work or home. We knew we'd see each other again in a week or two.

Giving any kind of service to others will give you a sense of purpose, a feeling that's like fresh ice water to your soul after it's been parched from the endless rat-race to self-service and self-glory...a feeling of contentment that you're fulfilling your duty to serve others. Take time—some time, any time—to serve others. It's a wonderful example for your kids. And when in doubt about how to do it best, just remember the Golden Rule: *Do Unto Others as You Would Have Them Do Unto You.*

"What we have done for ourselves alone dies with us; what we have done for others and the world remains and is immortal."
– Albert Pike -attorney, soldier, writer, and Freemason

CHAPTER 13

"Death Does Not End It"

The Grand Opportunity in Adversity

Good men must die, but death cannot kill their names.
— Proverb

"Sergeant Klein!" the voice cracked through the canvas tent late one evening. "Major Broderick needs to see you."

I walked out of the tent—we were back in Kuwait after being in Iraq and were waiting for our orders back to the States—and across the hard sand, heading to where the senior staff was bunked. "Sergeant Klein reporting as ordered, sir," I said as soon as I got there.

"Sergeant Klein, we need to head down to the Red Cross tent," he said. He sat down on the ammo box at the end of his rack to tie his boots. "There's some news from home."

That wasn't what I wanted to hear. No one wants to hear from the Red Cross when you're in the military—it usually isn't good news.

Without saying a word, I nervously walked with the Major down the sloping sand hill, looking out over the Kuwait City sunset skyline across the bay as we went. The spear-like tips of the Kuwait City Towers looked like they had impaled a few falling stars the night before.

When we arrived at the small Red Cross tent, it was empty; perhaps the clerk had momentarily stepped out for a head call or some dinner. A yellow legal pad lay open on the desk near the phone, though, so I read it:

Elizabeth Klein

Sergeant E. Max Klein-Mother

Cancer-Abdomen

Prognosis-Poor

Life Expectancy-Poor

My face went cold, and I felt like I'd been slugged in the gut. My heart raced. The next few minutes were a blur.

"You can call home from our phone here," said the clerk as he walked back in.

My dad answered.

"They've been doing some tests. I hear they may send you home early. She's going to have an operation at Johns Hopkins in Baltimore around Mother's Day."

A lot of things go through your head when you hear news like that. It was a surreal state of affairs—just as the last few months in Iraq had been—but much, much worse. People dying who you don't know is one thing, when you know them and love them it is completely different.

A few days later, I was in a United Arab Emirates airport, being detained by security while they tried to decide why I was traveling alone back to the States with no luggage at a time when most American military personnel were heading the other direction. The Red Cross letter explained things to them. I may very well have been one of the first servicemen in Operation Iraqi Freedom to fly back to the states on a commercial flight, actually.

When I arrived in Atlanta hub, I was wearing the civilian clothes that had spent the last few months in a bag. It felt strange to be wearing them again.

It's hard to describe the feeling of being back in the United States after you have been somewhere like Iraq. It felt safe; I felt thankful to belong there. It also felt somehow protected the rest of the world— insulated, less vulnerable, and warm. A good place to have children. It felt like freedom.

It would have felt peaceful, too, if it weren't for the turmoil dwelling inside me.

> The ultimate measure of a man is not where he stands in moments of comfort and convenience, but where he stands at times of challenge and controversy.
> – Dr. Martin Luther King, Jr.

Kuwait City skyline at dusk from our Marine Base in Kuwait

Adversity doesn't only come in the form of death, but death is usually the worst form of adversity. If we can handle that, we can handle anything.

Adversity is defined as "adverse fortune or fate; a condition marked by misfortune, calamity, or distress." Adversity can destroy you or strengthen you. For many, faith in God determines what effect it has. Adversity is the occasional hailstorm in life that bombards the inner oak tree of honor that you've been growing in your heart. If your tree is still young, the hailstones of adversity may destroy it—but as your tree gains strength, adversity will only disturb it a bit. In time, the hail of adversity will melt and water your honor's roots and give it even greater strength.

The secret of the wise is that they gain strength from adversity. They move forward from it, and they continue to nurture their trees of honor so the trees can eventually grow leaves that are capable of sheltering new "plants"—children—from the hailstorm of whatever adversity might come their way. That kind of strength (and ability to learn and grow) from adversity is one of the things that makes a good parent a good parent.

> Tell me not, in mournful numbers,
> Life is but an empty dream!
> For the soul is dead that slumbers,
> and things are not what they seem.
>
> Life is real! Life is earnest!
> And the grave is not its goal;
> Dust thou art; to dust returnest,
> Was not spoken of the soul.
>
> – Henry Wadsworth Longfellow

My mom was dying of cancer—month by month and week by week, she got weaker. When she couldn't walk any more, my four year-old cousin came to visit

My mom was sitting on the couch in her bathrobe, her body a frail shell of what it used to be...yet that ever-present warmth I remembered from my earliest infancy still radiated strongly from her. The tree of honor in her heart was the biggest in the family forest—it protected us all. Now that her body had become smaller, she had more room for the light of life. That hadn't weakened at all—in fact, sometimes she looked brighter than other people did, more radiant.

As my mom sat on the couch with a peaceful smile on her face and a gentle grace in her diminished body, my cousin pulled her mom—my aunt—aside and whispered in hushed awe, "Mommy, she's so beautiful!"

Children can see God's presence more clearly than adults can sometimes.

Even in the midst of her adversity, my mom was more concerned about us than herself; the weaker her body became, the more her dignity and love and honor showed. The way she handled her adversity—and the way my father handled it, with strength, love, and non-stop companionship—speaks of parent I want to be and of the person I want to be.

The display of dignity that seems to always be one step ahead of the adversity it encounters is called **Grace**. As a parent, you are always an example, even when your child is grown. I was watching both my mom and dad during this time, and their graceful example led me through our communal adversity. This doesn't mean there's no confusion or crying or pain or difficulty, it means meeting these things with love, gracefully. My parents having faith-fortified grace— their actions made it clear—, was the best way to deal with adversity, and nurturing a healthy tree of honor means that our hearts will always be lush with the leaves of grace.

Your children will watch how you handle bad things—not just death, but unemployment, accidents, all of life's adversities. Err on the side of love. Look to any bright side there is and always strive to keep humor and goodness handy and never give up hope. Establish a mood of strength and comfort. If you do all of this, you'll always be one step ahead of adversity, you will set the leadership example of *grace*.

Your children's souls are watching you, and your leadership through adversity will strengthen generations that follow you.

"I rest in the hammock of God's love."
– my mom describing the peace that passes all understanding as she was facing a physical—not spiritual—death.

Her spirit is alive and well with me, and even with my son, who never had the chance to meet her. He is a happier child because of who I am…and I am who I am because of who she was. Her last wish for me was to seek and find God because she knew what that meant. Her trust in God through Jesus Christ made her the powerful example of grace and joy that she was. Faith is not only for strength in bad times, but for happiness and peace in good times as well. I can only ask that you give God a chance and give your kids a chance to seek him while you're at it; it may possibly be the greatest act of leadership you ever perform. The two simple words of *Trust God* could possibly make all the difference in your family's future as it has mine. I'll leave that up to you.

> "A man can no more diminish God's glory by refusing to worship Him than a lunatic can put out the sun by scribbling the word, 'darkness' on the walls of his cell."
> – C. S. Lewis

CHAPTER 14

"From Dictator to Oil Boy"

The Fuel of Humor and Attitude

"Flush twice—it's a long way to the chow hall."
– writing on bathroom wall near the toilet at one of Fort
Indiantown Gap's white-wood-sided barracks in Pennsylvania

"I'll give them something to film!" I told Sergeant Clarence Davis (the same guy from the chapter on principles), the Marine who sat next to me in the rickety observation tower. He was controlling fixed-wing aircraft (jets) while I was controlling rotary-wing aircraft (helicopters) during a training exercise with the Filipino Marines.

I clicked a button on the handset and told the CH 53 helicopter pilot to fly over the hilltop we were on prior to returning to base.

"Roger, inbound ETA 5 Mikes." (Estimated time of arrival: 5 minutes.) From where we were sitting, we could look out over a huge sandy valley called Crow Valley. That was where aircraft would drop training bombs and helicopter gunners would shoot at targets with their 50 caliber machine guns.

A few minutes ago, we'd seen what we assumed was a news crew trekking up the hill to videotape the work that our combat engineers were doing on the hilltop. (They were constructing a cinderblock building that was to be used as an observation post by the Filipino Marines doing training exercises in Crow Valley—a gesture of thanks and unity to our Filipino counterparts.)

I gazed down at the reporters and wondered why they weren't coming over to see the building. I set down the radio receiver. A few minutes later, I saw the helicopter coming through the valley and directly toward us. *What a shot!* I thought. *The cameramen are going to love this!* The helicopter got closer.

Thump,thump,thump,thump... The rotors of the helo popped over the hill. It hovered above us, battering us with exploding dust and its incredible sound. Cameras rolled. A minute later, the helo veered off down the side of the hill, into the valley, and back to base.

Perfect! I thought. *Great shot for them.*

As the dust settled, a young Public Relations lieutenant came sprinting toward our tower. His face was contorted with exasperation.

"What the hell are you doing, Sergeant?" he yelled.

"Sir, what do you mean? I was giving the news crew a good shot so they didn't waste a trip up here."

"Sergeant, that news crew is doing a story on how U.S. military exercises are disturbing the tranquility in Crow Valley!!!"

"Uhh, sir, uhh...I guess I proved their point," I said as I looked at Sergeant Davis. He couldn't stop laughing.

The disturbers of tranquility approach.
The news crew is near the checkered building.

Dealing with life—everything in life—requires a sense of humor if we're going to maintain our sanity even during adversity. Whenever we get the chance, it's best to incorporate humor into our lives—good humor, that is. Good humor is productive and genuine and does not involve *anything* being done at someone else's expense.

You should always be comfortable enough with yourself to let loose and have as much fun as possible and even make fun of yourself. Being a leader does not mean that you have to be serious all the time by any means. I can't tell you what's funny, because that depends on your sense of humor—it's enough to know that it's imperative to have a healthy, active, non-lazy sense of humor in order to be a well-rounded parent and leader. Humor and a good attitude can make anything bearable and life much more fun. Exercise your sense of humor any chance you get.

"A sense of humor is part of the art of leadership, of getting along with people, of getting things done."
– Dwight D. Eisenhower

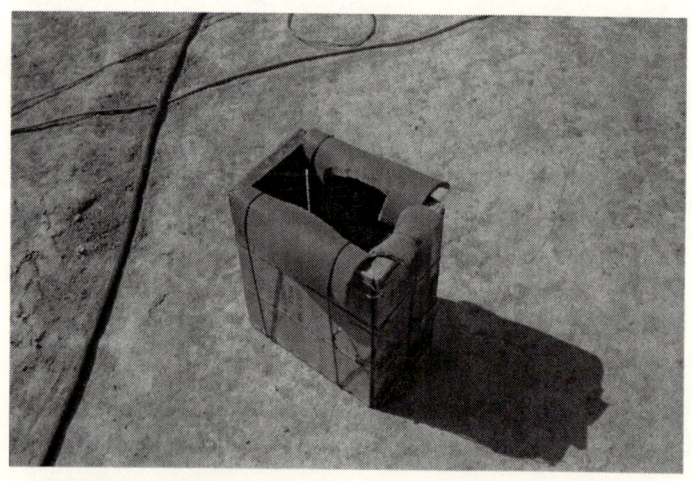

This fancy, padded ammo box "toilet" sold for two cans of chew *en route* to Baghdad; it sure beat "squatting".

Just a year after the episode with the Filipino news crew I sat alone and exhausted—in a black, dusty theater room. The theater occupied a central spot inside a compound in eastern Baghdad that was surrounded by 15-foot-high walls. We—the first Marine Regiment—had secured the area a few days before and set it up as the Combat Operations Center for the First Marine Division.

A few days before we took over, this same compound had been the Iraqi Secret Police Headquarters. Dust from our bombings (that someone yelled about in Chapter 2 "We're bombing Baghdad!") stuck to the seats and puffed up into clouds that rose on both sides when you sat in the chairs. We had seen more death on the way to where we were now and the hints of it were all over the compound. Upon further inspection of the buildings, we had found hundreds of files of information on prisoners that were held and tortured within these walls. We handed them over to the command to sort out. Civilians, whose relatives were probably in these files, lined up outside the walls hoping that they'd find some sort of information on their family members who had "disappeared" into these walls in the past.

Since combat operations were on the clock 24 hours a day, there had to be "shifts" of work. During some down time, I had walked into the theater and sat down. It looked like a typical American high school auditorium, with sloping aisles and a flat wooden stage in front. Near one side of the stage sat a large,

almost throne-like chair...probably a chair Saddam
Hussein and other high-level officers had sat in. Other
than that chair, the stage was empty. What had gone
on here? Was this the theater we had glimpsed on
television, the theater where Saddam had called out
the names of those officers whose loyalty had been in
question? Those men had been cuffed and escorted out
of the room, their faces white with impending death as
they exited the frame.

I noticed Marines in the theater were running
wires toward the center of the seated area. I sat down
in a dusty seat in the back and closed my eyes, half out
of fatigue and half because I wanted to imagine what
had happened here. I started to nod off with thoughts
of torture chambers and kidnappings and every other
terrible thing that had happened near this place
swimming through my mind, but suddenly, a noise
jolted me awake.

"Heyyyy....you forgot your briefcase!" yelled Jim
Carrey as he chased Lauren Holly through the airport
in his limo driver's outfit.

The communication Marines and generator operators
had converted the place into a working movie theater.

Thank God! I thought as I laughed at one of the
best movies ever made. *Just what I needed!*

I watched the whole "Dumb and Dumber" movie,
all the way to the end where Jim Carrey gives up his
chance of being an "oil boy" for the bikini team. Final-

ly, the credits rolled, and when they did, I felt refreshed, recharged, and in higher spirits.

I went back to my work. Baghdad was in a state of monumental change and we were its agents. And I was ready to go! It's amazing what honest laughter can do to recharge our souls. Teach your kids what good humor and a good attitude can do by demonstrating them as often as possible. And your sense of humor doesn't even have to be as highly sophisticated as mine to make a difference.

"When I was a boy of fourteen, my father was so ignorant I could hardly stand to have the old man around. But when I got to be twenty-one, I was astonished at how much the old man had learned in seven years."
– Mark Twain

The makeshift movie theater in Baghdad: Saddam center stage one day, Jim Carrey the next.

CHAPTER 15

"Padlocks and Persistence"

The Necessity to *Drive On*

"Nothing in the world can take the place of persistence.
Talent will not; nothing is more common than unsuccessful
men with talent. Genius will not; unrewarded genius is almost
a proverb. Education will not; the world is full of educated
derelicts. Persistence and determination alone are omnipotent.
The slogan 'press on' has solved and always will solve
the problems of the human race."
– Calvin Coolidge

My arms burned. We were all in push-up posi-
tions...I'd lost track of how long we'd been that way.

It was Boot Camp, and we were in the squad bay
late one evening. Just after we'd finished eating and
returned to the barracks, the drill instructor had
noticed that someone had left their footlocker unlocked.

Chaos ensued. We were ordered to remove all 67
locks from every foot locker in the entire squad bay
and hold them in front of us. The drill instructor
walked by and locked them to each other one by one.
(Keep in mind that they all looked exactly the same.)
Then we got into push-up position while one by one we

each tried the combination on every single lock until we each found the one that was ours.

If you don't have the right attitude in Boot Camp, this type of thing could drive you mad...and sometimes it does. But the point was three-fold: 1) always be accountable for your gear, 2) perform as a team under stress while controlling your emotions (either lose them or channel them productively), and 3) persist until the job is done, no matter how frustrating and seemingly meaningless it seems to be. I can't remember if we were in push-up position for the whole hour that it took to unlock the mass of locks, but I do remember that it was mind-numbingly frustrating experience...depending on how you looked at it. This type of thing day after day will bring out someone's true colors; some people snap, some threaten suicide, some think it's fun, and some just see it as what it is, mental training for more important things in the future. I was the last two.

These kind of things happened every day in Boot Camp and they began to change our attitude. We began to fully accept the things we can't control—bad things, frustrating things, things that seem impossible to fix or uncomfortable to fix, things that make you angry—but we must drive on, adapt to the situation, and keep going until the job is done and to NOT let our emotions interfere with performance unless it's improving it... and you know, after a few hundred Boot

Camp games like this, you realize your emotions can be a disruptive force in your life if they're not controlled and that any obstacle is surmountable with persistence.

> "Obstacles cannot crush me.
> Every obstacle yields to stern resolve.
> He who is fixed to a star does not change his mind."
> – Leonardo da Vinci

Persisting is one thing, but perfection at anything—including parenting—is an impossible goal. (Mark Twain once famously said, "I haven't a particle of confidence in a man who has no redeeming petty vices.") We've learned a lot so far about principles of leadership, but we've got to remember we'll never succeed at them all. Leadership is not about being perfect. If we truly thought achieving perfection was possible, we would drive ourselves crazy trying to achieve it. And if we do act "perfect" (perfect in our minds, at least), we will turn off those around us, including our kids—they know we're not perfect! It *is* possible, however, to persist in our self-improvement, to persist in working on the *fourteen leadership traits* and *eleven leadership principles* steadily through life.

Parents who persist despite obstacles, failures, and set-backs are real *parents*—they're neither stagnant nor are they perfect. If they don't persist in a somewhat regular effort to improve themselves, however, they'll become complacent in their ways and habits. Just remember that all we can do in life is give it our best shot—while we can't guarantee success in raising our kids, we can persist in improving ourselves and not worry about the rest.

And of course, we will occasionally fail along the way—anyone who has the courage to do things will fail sometimes—but failure is okay because it teaches us to *adapt*. And once we learn to adapt, we can learn to focus our efforts and *overcome* our failures. Then, instead of being defeated the way so many choose to be, simply because it's an easier path, we can learn to FIND A WAY to *adapt*, to *overcome* the problem, and to drive on. <u>*Persistence is required to fuel all the principles we've learned in this book*</u>

But be sure <u>not</u> to persist in correcting or perfecting yourself or your kids to the point of nagging and pickiness. Mind the important things and let the other things be. Take it easy. If we've instilled solid life basics in ourselves and our kids, everything else will eventually fall into place. You started your kids off on the right path, after all, and even though sometimes they might veer off it, they'll eventually come back to it. They are influenced by our broad, brushstrokes of real leadership, not by our irritating pencil stabs of nagging corrections

Day by day, bit by bit, by practicing leadership traits and pursuing honor with good humor and a good attitude, and not getting too serious about perfecting things, we *will* change ourselves and *that* will, someday, change the world.

> "Most of the important things in the world have been accomplished by people who have kept on trying when there seemed to be no hope at all."
> – Dale Carnegie

A few months after we'd sorted out our padlocks, we stood outside in the cool breaking sunrise. Our legs felt like jelly, our blistered feet burned, and our bodies were weak, but we stood and waited for what we'd been pushing to achieve every day and every night over the past 12 weeks. When we dropped the heavy packs from our backs, our shoulders felt cool as the sweat that had collected between our packs and our shoulders began to evaporate with the breeze.

It was the end of the twelfth week of Boot Camp on Parris Island. We had just finished the "Crucible," a three-day training exercise consisting of sleep deprivation, food deprivation, mental tests, and long marches. It was the final stage of the rite of passage called Boot Camp. We had been marching since 1AM and had finally arrived back, blistered and hungry, at the parade grounds.

This was it. We had finished many arduous weeks of training, and those who were still there had "earned the title" of United States Marine. We were about to receive the Eagle, Globe, and Anchor to signify that we were now Marines. (The Eagle, Globe, and Anchor is the emblem of the Marine Corps. The eagle stands for the United States, the globe stands for continuing worldwide service, and the anchor stands for the naval heritage of the Marines and the organization's existing relationship with the U.S. Navy.)

We stood at attention with our shoulders back. One by one, the drill instructors stood in front of each Marine, handed him a small black Eagle, Globe, and Anchor pin, and shook his hand.

"Welcome, Marine," they said as the "Marines' Hymn" played from a small CD player lying on the ground. Chills shot through our bodies and tears came to our eyes. We were Marines now! We had reached our goal by being PERSISTENT. I still get chills when I hear the Marines' Hymn...every time. And everything I've accomplished that required persistence has planted a similar seed of nostalgia in my heart. Persisting will change your life. And there's no doubt the persistence in Marine Corps training has changed the world, just as there's *no doubt* your persistence in parenting and personal leadership development will change the world, in some way, for the better.

"Some people spend an entire lifetime wondering if they've made a difference. The Marines don't have that problem."
– President Ronald Reagan.

And neither do you! You know the universal leadership traits and principles, now all that's left to do is to *develop them and persist in doing so.* Persist in learning and developing and leading your children by example as well as you know how, and forget about how you're doing in life compared to others. You don't have to be the best and brightest leader to make a difference, you have to try to be *your* best and *your* brightest because that is how you will make the biggest difference. And always remember to let your child know that you love them and they're good enough for you no matter what mistakes they make. In the wake of those failures, that will come to you and your kids, find any positive things you can and encourage, encourage, encourage; *this* will fuel persistence!

"If a man has done his best, what else is there?"
– General George S. Patton, Jr.

"I long to accomplish great and noble tasks, but it is my chief duty to accomplish humble tasks as though they were great and noble. The world is moved along, not only by the mighty shoves of its heroes, but also by the aggregate of the tiny pushes of each honest worker."
– Helen Keller

CHAPTER 16

"Adventure Everywhere"

Setting the Example of Wonder and Curiosity

"The most beautiful thing we can experience is the mysterious. It is the source of all true art and all science. He to whom this emotion is a stranger, who can no longer pause to wonder and stand rapt in awe, is as good as dead: his eyes are closed."
– Einstein

I stopped walking and listened with all my senses. Somewhere, a bird chirped. Its voice echoed through the palm trees and out into the open expanse of the ocean. Waves crashed rhythmically onto the black rocks.

I opened my eyes and looked around. If I'd been looking at the same landscape sixty years ago, or a hundred years ago, or a thousand years ago, I would have seen the same scene...except for one thing. That one thing was the cement wall that protruded from behind the creeping vines.

The only thing that had changed about the wall since it had been built was that it had been overgrown with vines and moss. Aside from its green carapace, though, the wall looked just as it must have when the Japanese had built the wall as part of their fortifications for the defense of Tinian Island in the Marianas Islands during World War II.

The wall wasn't roped off. Herds of picture-snapping tourists weren't photographing themselves in front of the wall to prove to their friends that they'd "been there." There was just the wall, the jungle, and the ocean.

What did the Japanese think when they saw the U.S. Navy boat come over the crest of the horizon? I wondered. *What was going through their minds when diesel exhaust from U.S. Marine landing crafts blanketed this wall? What happened right here where I'm standing? Just imagine…*

I walked around the bunker down a long path and came into a clearing. A wooden sign stood next to small patch of grass: "Bomb Pit #1."

A nearby plaque read:

"From this loading pit, the first atomic bomb ever to be used in combat was loaded aboard a B-29 Aircraft and dropped on Hiroshima, Japan, August 6th, 1945." (One year after Private First Class Wilson died there and won the Medal of Honor in the battle to take Tinian Island) (see chapter 9)

The world had been changed forever from something that had taken place right under my feet. And this was only possible in part because of the courage and leadership of PFC Wilson. All things in history are tied together and goodness or evil in history depends on the sum of the leadership traits and principles of individuals living at that time; it's the same now. The future is completely dependent of the cumulative efforts of us, the leaders. Other than a cataclysmic natural disaster, parenting and leadership is the only thing that will define the future. **That's why this all matters.**

I closed my eyes, breathed in the clean ocean air, listened to the rustle of the trees. Time and imagination, I reflected, are all that separate us from witnessing the greatest events in history. If we can develop our imagination and draw inspiration from the fixed sensations of our surroundings, we can almost feel like we were there.

> "One thing that life has taught me: if you are interested, you never have to look for new interests. They come to you....all you need to do is to be curious, receptive, eager for experience. And there's one strange thing: when you are genuinely interested in one thing, it will always lead to something else."
> – Eleanor Roosevelt

Driving through the beautiful island of Tinian

"In the hopes of reaching the moon, men fail
to see the flowers that blossom at their feet."
– Albert Schweitzer

There's a whole wide world of amazing and fascinating things under our noses, but we must get off the
beaten path and out of our daily ruts of habit to see
them. As we become more confident and competent
leaders and *know ourselves* better from our well-
rounded education, we understand our lives better and
start to become genuinely fascinated with the rest of
the world and its people. In the military, very often
soldiers have free time to explore the cultures and
natures of the countries in which we find ourselves.
Soldiers aren't tourists—we have the opportunity to
experience cultures in an untraditional way, and we're

encouraged to understand the "hearts and minds" of the local people. We're taught to respect them.

As soon as we realize that traveling is more than pre-packaged tours and photo snapping, we can all experience the world this way. There's a time and a place for pre-packaged adventure, yes, and I've been on some great guided tours myself, but in many cases, the same experiences a tourist gets from a bus tour could be had by watching a travel show on television. The excitement, the adventure, the sights and smells and uninterrupted absorption of the *essence* of a place only happen when we're really *in* that place!

Once we've felt that sense of adventure, we want it again and again. We recognize that there's something new around every tree, in every village, in every conversation with the locals; just as there's something new in every new experience we have the courage to embark on.

The military can thrust us into that true adventure travel...but we can also get it on our own. Avoid the beaten path! Be unscripted in your travel. Over-planning results in stresses that blind us to unforeseen benefits. Travel doesn't have to be international, either—it can be right outside your front door or within the pages of a good book.

Start young—go on "adventures" in the backyard with your toddler. Boredom is the phlegm in the lungs of life, and it takes action to get it out of your system.

Lead by Example and wonder about something, talk about something, work on

something...but always do *something*. Of course, sometimes you just need to relax, and there's a time for that, too. But be the active example that keeps boredom from slithering slowly into your child's mind and crawling into their spirit.

When people are talking, listen. Listen to their conversations and ask questions. The rich diversity of other cultures is fascinating. Don't view what you're experiencing through the lens of your former percep-tions—imagine that you *are* the other person, living in his or her lifestyles and in his nation. With art, try to feel what the artist *meant* when he or she created it.

The world of nature is always fascinating. Smell the rain, feel the trees, touch the rocks—physically absorb what's around you and put yourself into the moment.

Your daily life will become much richer when you realize there's a world of wonder waiting to be savored after you peel away the dull crust of daily habit. Curiosity is the key to unlocking wonder, and wonder will lead you to the road to adventure. You can start doing this at any age—there's always time to expand our horizons.

If you can be curious, and open new doors in your life, you'll find that you're a little different afterward, a little bit better as a leader and a parent. Adventure is everywhere—we just have to know how to find it.

"I think, at a child's birth, if a mother could
ask a fairy godmother to endow it with the most useful gift,
that gift would be curiosity."
– Eleanor Roosevelt.

Wonder and curiosity also leads to unity with your fellow man.

A few months after I saw the wall, I found myself sitting underneath a ragged umbrella, on a dirt street leading to the Cambodian border. It was a hot, muggy evening; other people had gathered underneath the umbrellas alongside us to sip cool drinks.

A Thai Marine bought Singha beers and shots of rice liquor for a few of us and told us about his life. His wife had died and most of his family was gone, but he had his son with him—a happy little six-year-old boy who was getting us cold beer and drinks.

The Thai Marine told us what it was like to grow up in his country and then asked us questions about ours. When you talk to people long enough, you realize they all want the same thing: security and happiness for their children. He seemed to be providing his son with both. Their clothes were dirty and their pockets were empty, but they were together and they were happy.

Curiosity and wonder can make your whole life an adventure—there's so much to learn.

"Wonder is the beginning of wisdom."
– Greek proverb

CHAPTER 17

"From Baghdad to Babies"

The Hope For Us All

"Why we care"

This little Iraqi girl's future grandchildren will care how we raised our kids, because as parents, the world depends on all of us to be leaders. Our efforts span generations and oceans and deserts—in some small way, what we do with our kids will affect what happens in every corner of the globe. What we do today will change the world forever.

Big, beautiful, heavy-looking snowflakes fell softly onto the Pennsylvania fields. It was the kind of snow that muffles all outdoor noises to about half their normal volume and makes a silky-light crunching sound when you take a step on it.

I looked out through the maternity ward window and wondered how our inbound families were faring on the roads. My wife's contractions had started, but they were still happening far apart; she had a while to go.

People came in during their lunches and after they'd left work. A sense of surreal excitement buzzed around us. Despite that, though, my wife was getting tired, so as soon as the last visitor left, she took a nap. I rested on the couch and thought about the new son who was almost ready to meet us. The love I had for him had begun the day I found out he was growing in her womb. The kind of love a parent has is a love that's unexplainably strong, a love that's more pure than even marital love. A love based on complete creation and dependence—like God's love, it's unconditional, pure, true love.

Her pains started coming on stronger at midnight, but fortunately, the medication allowed her to get a few more hours of sleep. When the sun came up the next day, the snow had finished falling. We waited.

A gradually-building flurry of nurses woke up the room, and after a burst of activity, I found myself at my wife's side. It was one of those times in life that felt like it had its own reality, where I saw everything differently somehow...a more-real time, a time when the normal habits, thoughts, and attitudes that pile themselves into everyday life were no longer important. It's pure, un-filtered humanity, and it makes you feel *alive*!

She pushed and pushed and pushed some more. I saw my son's head, and then his face, and then his shoulders, and then he was here! The nurses quickly took him to the warm table where they cleaned and wrapped him. They immediately wanted to get him to the machines to monitor him since he was five weeks early, but before they took him out of the room, they said to my wife, "Do you want to kiss him?"

"Yes," she whispered, with tears in her eyes. His tiny eyes were open and he was crying, but when my wife kissed him on the head, his crying stopped for a second. We both just stared at him, caught up in true love.

Nine months later, I see him lying on the bed in his diapers, laughing a contagious and hearty belly laugh as my wife sweeps her hair across his bare belly. His fingers grasp at her hair, strong and sure with innocent happiness. This is the peace that we live to enjoy; it's the hope for the future that we're willing to die to protect.

A beautiful palm grove just east of Baghdad.

And I stripped off my ragged clothes and stood in the light of the hot sun, pausing for a second before I stepped into the shower and turned on the tap to let out the cool river water that had been pumped from the Euphrates River. It was the same water that had once cradled the first civilized human beings the world had ever seen...and now it trickled over my head and down my body, a refreshing reminder of life in a scorching desert.

There is beauty even in a war zone, I'd discovered. I'd seen death close up and destruction everywhere and I'd witnessed adversity, but just like in the rest of the world and in all of our lives, there is also beauty everywhere. Leaders can see it! That beauty is even more powerful than the destruction and even more visible if we know where to look—we just have to *take the responsibility* for developing an eye for it.

The cool water falling from the makeshift military showerhead quenched my dry, thirsty skin. A feeling of satisfaction and contentment came over me. Besides feeling refreshed from taking my first shower in 32 days, I felt something else, something much deeper...something that made me feel more alive. It was that feeling of having made a *good* difference in history—a clear, calm, contentment and a clear sense of purpose. Life felt as black-and-white and as peaceful as it had when I was a child.

Later, I had that same feeling when my child was born, and I still have hints of it when I teach him something new. We can all have that feeling when we realize we've taught our children something real and something good. When we teach our children how to live a good life by setting the example and by leading them, we strengthen their honor and light their souls on fire. That's a gift they'll take with them into their futures and a gift they will someday give to their own children.

So, as our days go by and our children grow, let's remember to occasionally take time to look around, and truly appreciate this gift we have of sweet, wonderful LIFE, and for the grand opportunity we have as leaders to make a difference! It's never too late for us to change ourselves. Life is WONDERFUL if we see it that way. No matter what happens to us...*life*, even a tough one, is better than the alternative.

My son is growing now and I see goodness in him. I've been called many good things over the years like Husband, Son, Grandson, Friend, Sergeant, Marine, Graduate, Recruit, American, but the greatest thing anyone has ever called me by far...is Daddy. No other title has evoked the emotion, the sense of responsibility, the love...or the JOY that I feel when I hear my son call me that. I have a duty to live up to that title...an exciting opportunity to live up to it!

As parents, if we've put food on our table and we love our kids, we have lived up to that duty and we're already doing a wonderful job, but now we also know the tools of leadership which help us be even better; the <u>14 leadership traits</u> and the <u>11 leadership principles.</u> Adding these to our lives will be the icing that makes our lives so sweet. This book was written to strengthen these principles in you, the parent and leader; because I know that they produce positive results that will improve your life, your children's lives and everyone's lives around you.

These traits and principles are yours now, so maintain them as you wish and put your own style into them. I wish you and your family the best. The stories you've read are just my stories, you'll be making far more interesting ones every day until the day you draw your last breath, and long, long after that.

Someday, because of your efforts today, your children will look back and be so thankful for the fact that you lived as a leader of character. The world will be a safer, your family will be happier, and you will have made a difference that will be felt for the rest of time, just because of who *you* chose to be. LIFE as a loving leader...ahhhhh...what a wonderful and endless opportunity; and *you* have already seized it!

"Start children off on the way they should go, and even when they are old they will not turn from it."
– Proverbs 22:6

"From all of us throughout the world and all of us yet to come, Thank You!

<u>Appendix</u>

SUGGESTED READING:

1) THE BIBLE (EVERYTHING YOU NEED TO KNOW)

2) HOW TO WIN FRIENDS AND INFLUENCE PEOPLE by Dale Carnegie (DEALING WITH PEOPLE)

3) TUESDAYS WITH MORRIE by Mitch Albom (EMPATHY AND DEATH)

4) GEORGE FOREMAN'S "FATHERHOOD BY GEORGE" (CHRISTIAN FATHERING PRINCIPLES)

5) THE 5000 YEAR LEAP by W. Skousen (AMERICA'S EXCEPTIONALISM/FREEDOM)

6) Band of Brothers: E Company, 506th Regiment, 101st Airborne from Normandy to Hitler's Eagle's Nest by Stephen E. Ambrose (BOOK OR MOVIE) (RESPECT FOR WHY WE HAVE WHAT WE DO, AND GREAT EXAMPLES OF COURAGE AND ADVERSITY)

7) THE PURPOSE DRIVEN LIFE by Rick Warren (HELP YOU FIGURE OUT WHAT YOU WERE MEANT TO BE IN LIFE)

8) Anything by James Michener (HISTORICAL FICTION THAT JUST MAKES THE WORLD AND ITS PEOPLE AND PLACES MORE INTERESTING)

Footnotes

1. JOHN BROWN'S WORDS FROM CHAPTER 1:
    Article written soon after execution by David
    Hunter Strother (Porte Crayon). The article was
    intended for publication by Harper's Weekly, but
    because the material was deemed too controversial,
    it wasn't published until 95 years later, when Boyd
    B. Stutler's article "An eyewitness describes the
    hanging of John Brown" was published on Ameri-
    canHeritage.com.
2. Rhodes, James Ford (1892). *History of the United States
    from the Compromise of 1850*. Original from Harvard
    University: Harper & Brothers. pp. 385

LEADERSHIP TRAITS-JJ DID TIE BUCKLE

(WALLET OR WALL CUT-OUT...OR SEE
WWW.LEADERSHIPMANUALS.COM
FOR A FREE PRINT-OUT VERSION)

JUSTICE is defined as the practice of being fair and consistent. A just person considers each aspect of a situation and then rewards or punishes accordingly.

JUDGMENT is the ability to think about things clearly, calmly, and in an orderly fashion so that you can make good decisions.

DEPENDABILITY means that you can be relied upon to perform your duties properly. It means that you can be trusted to complete a job. It is the willing and voluntary support of the policies and orders of the chain of command. Dependability also means consistently putting forth your best efforts in an attempt to achieve the highest standards of performance.

INITIATIVE is taking action even though you haven't been given orders. It means meeting new and unexpected situations with prompt action. It includes being resourceful when something needs to be done and the normal material or methods are not available to you.

DECISIVENESS means that you are able to make good decisions without delay. Get all the facts and weigh them against each other. By acting calmly and quickly, you should arrive at a sound decision. You announce your decisions in a clear, firm, professional manner.

TACT means that you can deal with people in a manner that will maintain good relations and avoid problems. It means that you are polite, calm, and firm.

INTEGRITY means that you are honest and truthful in what you say or do. You put honesty, sense of duty, and sound moral principles above all else.

ENDURANCE is mental and physical stamina. It is measured by your ability to withstand pain, fatigue, stress, and hardship. Example: enduring pain during a conditioning march in order to improve stamina is crucial in the development of leadership.

BEARING is the way you conduct and carry yourself. Your manner should reflect alertness, competence, confidence, and control.

UNSELFISHNESS means that you avoid making yourself comfortable at the expense of others. Be considerate of others. Give credit to those who deserve it.

COURAGE is what allows you to remain calm in the face of fear. Moral courage means having the inner strength to stand up for what is right and to accept blame when something is your fault. Physical courage means that you can continue to function effectively when physical danger is present.

KNOWLEDGE is the understanding of a science or art. Knowledge means that you have acquired information and that you understand people. Your knowledge should be broad—in addition to knowing your job, you should know your unit's policies and keep up with current events.

LOYALTY means that you are devoted to your country, the Corps, and to your seniors, peers, and subordinates. The motto of our Corps is Semper Fidelis! (Always Faithful). You owe unwavering loyalty up and down the chain of command, to seniors, subordinates, and peers.

ENTHUSIASM is defined as a sincere interest and exuberance in the performance of your duties. If you are enthusiastic, you are optimistic, cheerful, and willing to accept the challenges. (USMC.MIL, 2003).

Marine Corps Leadership Principles

- Know yourself and seek self-improvement.
- Be technically and tactically proficient.
- Develop a sense of responsibility among your subordinates.
- Make sound and timely decisions.
- Set the example.
- Know your Marines and look out for their welfare.
- Keep your Marines informed.
- Seek responsibility and take responsibility for your actions.
- Ensure assigned tasks are understood, supervised, and accomplished.
- Train your Marines as a team.
- Employ your command in accordance with its capabilities.

About the Author

My mom probably would have wanted me to smile in this picture!
This was taken in Kuwait about a week before Operation Iraqi
Freedom started.

Max Klein is the father of the best kid he's ever met—
Cole! He's also the husband of Kamm, a wonderful
wife and spectacular mother. Prior to his service as a
Marine Corps Sergeant, he served in the PA Army
National Guard as a Field Artillery forward observer
(which meant that he got to go to Army Basic Training
and Marine Corps Boot Camp). He served with Marine
Air Support Squadrons 1, 2, 3, and 6 (where he at-
tended multiple leadership schools and courses and
performed as Master of Ceremonies for multiple 1000+
attendee events). He was attached to the First Marine
Regiment Headquarters Company in Iraq where he
was one of the first Westerners to enter the inside of
the Iraqi Secret Police Headquarters. During that

time, he was also a black-belt martial arts instructor-trainer for the Marine Corps Martial Arts Program (MCMAP).

He earned his MBA from Western Governor's University in Utah and now resides in Huntsdale, Pennsylvania with his family.

He is fascinated by the realization that understanding and applying just a few leadership skills can completely change people's lives. Because of his leadership experience and his desire to make life more fulfilling for others, he wrote this book on leadership parenting in the hopes that he could help fellow parents have an easier time of raising their kids to be honorable and life-loving human beings. This effort, in the end, will change the world and this is not as difficult as we may think; All we need is to do is apply a few leadership principles to our daily lives!

Breinigsville, PA USA
31 January 2010
231646BV00001B/1/P